"I do the very best I can. I mean to keep going. If the end brings me out all right, then what is said against me won't matter. If I'm wrong, ten angels swearing I was right won't make a difference."

Abraham Lincoln

"Joe Bastardi's love for the weather and climate drove him to write *The Climate Chronicles*, an exposé of the true climate change agenda. By drawing on many of the over 150 blogs he has written on the matter, he shows how weather and climate are being weaponized, politicized and in reality distorted by academia, media and even religious leaders to advance a cause that is counter to our nation's best interests. A must-read for anyone who loves weather and climate, cherishes the freedoms that are being attacked in our country today, and is curious enough to question what they are being told."

Sean Hannity
Fox News & Radio Talk-Show Host

"In all of the sound and fury surrounding the climate change debate, two things have been glaringly absent: rationality and truth. Renowned meteorologist Joe Bastardi restores both to the climate discussion as he methodically lays out actual science, places it in context and strips out the globalist and statist agendas that have poisoned the conversation from the start. This is a must-read for anyone interested in arming themselves with the truth."

Monica Crowley
Senior Fellow, London Center for Policy Research

"I was trained as an analytical chemist and medical doctor and I have been dismayed by how the globalists and their 'puppets' in the 'Weather Ambulance Chaser' contingent have denigrated the scientific method in their pursuit of societal control and the almighty dollar. Joe Bastardi is a dedicated scientist who utilizes the scientific facts and irrefutable data to set the record straight on why the climate has changed, is changing and will continue to change. He has created a 'Weapon of Mass Instruction' with *The Climate Chronicles*."

David H Janda M.D.
Orthopedic Surgeon

"Joe Bastardi hits home with this exposure of the political climate change myth. Exposing the Mikhail Gorbachev/George H. W. Bush agenda of 'Green replaces Red' at the end of the Cold War shoots a bullet dead center in the so-called global warming lie."

Craig Roberts
Editor-in-Chief for Consolidated Press Int.
Author of "The Medusa File: Crimes and Coverups of the U.S. Government".

The Climate Chronicles

Inconvenient Revelations You Won't
Hear From Al Gore — And Others

By

Joe Bastardi

An Original Publication of Relentless Thunder Press

Copyright ©Joe Bastardi

Cover Art ©Joe Bastardi

ISBN-13: 978-1984371409

ISBN-10: 1984371401

First Relentless Thunder Press printing: Feburary 2018

10 9 8 7 6 5 4 3

Relentless Thunder Press

Printed in U.S.A.

Acknowledgements

I can mention countless people in my life to whom I am indebted — debts that no honest man can repay. But in the end, it comes down to God, family and country.

My heavenly Father for the blessings He has poured out on me.

My family for providing a foundation for me to reach for tomorrow.

And my country for engendering liberty. Despite some faults, there is a comparison to the freedom and opportunity we enjoy because of those who fought to instill and protect them.

I believe we are in a great conflict. If open-minded people look with *gratitude* at what it took to get us here, they will realize that the trials we face are because we have been so blessed. If not for the advances made possible by fossil fuels, the people trying to destroy this important source of energy would not even have the means to do so. The same goes with the foundation upon which our nation was built. You must have gratitude above all.

If we have gratitude, we will understand that man advances through the challenges that occur from conflict via resolution and adaptation. It may not be what you believe, but it's what I believe. Nature does too. The hot summer day with the thunderstorm is cooled by that thunderstorm, and hours later the sky has cleared.

It's man who creates the distance. It's God who builds the bridge!

I am thankful for that providential bridge builder and for all that was created around me. In a way, I'm also thankful for my adversaries, because they reveal conflicts that need resolving. If we can accomplish that, then we are all better off.

I also wish to thank those who have helped with the production of this book: Jordan Payne, my editor who has the torturous job of polishing what I write in the opinion blogs; Michael Barak, who has suffered through my endeavors and who has advised in how to "brew the perfect cup of coffee" when opining on issues in the minefield of climate change; and Craig Roberts, an author in his own right who did final formatting and proofreading. All were instrumental in seeing this project through.

But most of all, thanks to my wife, Jess, for her support.

Introduction
To the Weather, With Love

This is a love story. I have loved the weather since my first memory. Next to my mom and dad, the weather and climate have been what I have loved the longest. It has given me a chance to make a living doing what I love. I see the majesty of Creation in action every day and, as such, it has drawn me closer to God.

This book is not meant to be an attack on others but rather a defense of what I dearly love. All these are opinion blogs, already published and organized in a way to tell this story. I sense that weather and climate are being used as a means to an end, and in a way, this ruins them. Would you not defend what you love? Would you not speak the truth as you see it? I am nothing more and nothing less than someone who is speaking up for what I cherish. This is what *The Climate Chronicles: Inconvenient Revelations You Won't Hear From Al Gore—And Others* is about — opinion blogs I have written to defend what I hold near and dear.

Furthermore, my purpose is to protect the longstanding forecast techniques that can help supplement the great advances in science. The argument comes down to a group of people who are trying to get the public to believe that every major weather event is a sign of climate change versus people like me who *use the past* and the techniques of today to forecast these events — almost always before agenda-driven Climate Ambulance Chasers even see them! (I use that term with some tongue in cheek, but it does remind me of someone who will use something adverse to profit under the guise of "helping.") It boils down to a defense of the absolute truth of nature's ways versus a relativist truth that depends solely on a missive and the desire for control.

When forecasting, I always try to explain *why* something will happen. I have found that in the climate debate, I am dealing with climate alarmists who do not make a forecast for a specific event. But if an event occurs that satisfies their longer-term abstract missive, they turn around and then claim it's because of what they are fussing about. In other words, any answer they can use for their ideas is correct. There is no wrong answer, and there is no time limit. For example: If climate alarmists say it will warm and it fails to, it simply means it's coming later. It's a deceptive process. Moreover, the very methods that are being used are projected dangerously on their opponents. Slanderous comments — like accusing people with whom they disagree of being "deniers" of basic science — are right out of Saul Alinsky's *Rules for Radicals*. Basic foundational science — such as using the past as a predictor of the future, or even questioning the events of today — is dismissed as useless by people pushing this.

Bill Nye, a prominent believer in anthropogenic global warming, revealed a lot when he alluded to the fact that the people who are resisting the so-called green movement are mostly older people and will soon die-off. He is correct, and in fact I wrote about it in "Bill Nye: The Real Message We Should Pay Attention To," published on Aug. 4, 2017:

> There was a minor uproar over a recent Bill Nye comment that is summed up in a Daily Caller article titled: "Bill Nye: Climate Change Scientists Need To Wait For Older People To Die."
>
> Let's look at this for what it really reveals.
>
> First of all, Bill is stating a fact. Many in the climate change "resistance" are Bill's age and older. But this generation was brought up differently than the current group of people, many rising through academia. We were taught to *question authority*. We were also

encouraged to reject groupthink. Perhaps it had to do with Dwight Eisenhower's farewell speech in which he warned against the military-industrial complex — which, when I was growing up in the '60s and '70s, was being used as a rallying cry against our involvement in Vietnam. But those of us who are in the generation of which Bill speaks also took note of the other part of Eisenhower's speech.

Let me borrow from *Wikipedia*, where its note on the legacy of the speech attests to my assertion about its importance in relation to Vietnam:

"Although it was much broader, Eisenhower's speech is remembered primarily for its reference to the military-industrial complex. The phrase gained acceptance during the Vietnam War era and 21st-century commentators have expressed the opinion that a number of the fears raised in his speech have come true."

Here's that part of the speech:

"In the councils of government, we must guard against the acquisition of unwarranted influence, whether sought or unsought, by the military-industrial complex. The potential for the disastrous rise of misplaced power exists and will persist. We must never let the weight of this combination endanger our liberties or democratic processes. We should take nothing for granted. Only an alert and knowledgeable citizenry can compel the proper meshing of the huge industrial and military machinery of defense with our peaceful methods and goals, so that security and liberty may prosper together."

Here is one of the greatest generals of our nation warning against the military-industrial complex, and many took it to heart.

But we of that generation also knew about the second part of his warning.

"Akin to, and largely responsible for the sweeping changes in our industrial-military posture, has been the technological revolution during recent decades.

"In this revolution, research has become central. It also becomes more formalized, complex, and costly. A steadily increasing share is conducted for, by, or at the direction of, the Federal government. ...

"The prospect of domination of the nation's scholars by Federal employment, project allocation, and the power of money is ever present and is gravely to be regarded.

"Yet in holding scientific discovery in respect, as we should, we must also be alert to the equal and opposite danger that public policy could itself become the captive of a scientific-technological elite."

Ike was right.

Here is where Nye has a point. He understands that the people who were brought up in the form of Americanism and who believe that the individual should question authority are getting older and will not be around when the new vanguard takes over. Enlightened he believes himself to be, and I suspect others like Al Gore think they are simply leading the new wave to replace the old wave. But instead of attacking Nye and making it seem like he has a death wish for his opponents, why don't people actually look at the facts of what he is saying and what they actually mean for things like critical thought and skepticism? Those things are essential not only to the scientific method but also for the basis for man to use his free will to better himself. The bottom line is that Nye's statement does not identify a problem with Nye, it identifies a problem with what has happened over the post-Vietnam generations. Nye and the climate issue, when looked at more deeply, reveal a deeper problem that strikes at the core of what has led to the nation's rising.

Nye is right about the inevitable result. It does not make him right about CO_2 being the climate control knob. But Nye is not the demon here; he is more a messenger of the very changes that Eisenhower warned us about in his speech. And what is apparent is that the generation which followed that speech took his word to heart over one thing but went the opposite way on another.

Some may be tempted to think I am going soft on Nye. I am evaluating what he said in an objective manner. I would suggest that instead of tearing at the messenger we look at the message. For in his message is the real danger — not to the people who are aging but to the methodology that makes solid foundational skepticism and freedom possible.

I harbor no illusions about how this will end. It does not matter whether I am right or wrong about the climate, because the answer does not matter. Regardless of what happens, my opponents will claim they are correct and they have a willing media that refuse to look deeper into the matter. Additionally, while they accuse people like me of being funded by fossil fuel interests, the overwhelming amount of government and industry money is going to environmental activists and lobbyists. They're like banks now — too big to fail. Try getting a grant to do work that proves the other side of the issue. The truth is, you don't need to; the evidence is easy to find if you look. And the truth also is that this is not about stopping calamity.

For me, it's far simpler than that. It's about getting to the actual truth. If I were in it for the money, I would simply join the other side given the prevailing winds. That is where the money is. But it's not about that to me. The fact is that I have always loved weather and climate and I've used the past as the foundation I stand on today to reach for the future. Because of that, I am well aware of what happened before. It's a matter of saying something if I see something, and all I see here is a missive that is simply using what I love as a means to an end. And that end is politically, socially and economically driven. The idea that the means justify the ends makes it impossible for them to turn back now.

This book is a partial compilation of my writings at *The Patriot Post*, which has been an outlet for me to show how history combats hysteria. You will find that climate and weather go hand in hand, because the foundation you stand on today to reach for tomorrow was built by the past. Therefore, you need to know the past.

I have written about 150 opinion pieces on this matter and have consolidated some of them here with some minor editing/appending where appropriate. I also expound on many of the ideas I address in those pieces.

It's sad, you know. I work like a dog to nail a forecast for anyone. In fact, in the solar and wind industries — a profit center for me because not only do they need longer-term forecasts but day-to-day forecasts — accurate outlooks would bring down costs and help the industries become more viable. So it doesn't matter to me — just nail the forecast and find the best way to communicate the scoop via media, no matter the outlet. When you take out all the agendas, you can a) find the answer and b) do the best job possible when the best matters.

There is an assault on what I love. So, naturally, I will stand up for what I know is right. The fight is not to save the planet. The fight (if one wants to call it that) is simply one in which someone is doing something that I don't believe is right. If you were in this position, would you just let it go, or would you say something? I chose the latter, and that is what this compilation is — a defense of what I dearly love.

As I wrote on July 3, 2014, in "For the Love of the Weather":

> I will be in Las Vegas for the 9th International Conference on Climate Change July 7-9. I enjoy attending these because everyone I come in contact with is someone I can learn from. My father and mother taught me to surround myself with people I could look up to. I adapted that attitude in my training, in my work, and in the person I married. And when I go to these events, I am in awe of the people I get to talk to. I bring some of them up quite often in these pages.
>
> Given the situation we're in today, we all must be careful to not let our reasons for doing what we do go astray. I often feel bad for the people in the anthropogenic global warming camp. How can they possibly walk back their position? It's the reason why no matter which way a metric goes, they either ignore it or

conjure up another reason for it happening — because they did not forecast it to begin with. But what else can they do? Look at the people in their camp who have started to disagree. They are chastised and demeaned. There is a simple reason: The climate alarmists' goal, whatever it is — be it self-esteem, money, power, control, or all of the above — is their god. It's what they are forever in pursuit of. And when something becomes your god, you cannot defy it.

I constantly analyze anything that is important to me. (I drive my wife nuts. If I have a bad workout, or get a cold, I have to figure out why it happened.) And I always analyze my motives. For years I felt like Harold Abrahams in "Chariots of Fire." This line hits me hard because I know exactly how he feels:

"I'm forever in pursuit and I don't even know what I am chasing."

I try to be more like Pastor Eric Liddle, the other character of prime importance in the movie. To paraphrase: God made me for a purpose, but He also gave me a love for the weather. And I see His majesty in it every day.

What strikes me most about the Heartland Conference is that I am with people who are in love with the weather, the climate and their country, and many of them have loved these things longer and stronger than I have.

My stance on global warming is a product of my love for the weather. There is no goal. It's about having another chance to do what I was made to do. And when I'm with people who I sense have the same ideas, it somehow makes me stronger and more able to run toward what I was made for.

As one gets older, one can get tired. But only when your heart gives out does your strength give in. For me, all this is an affair of the heart.

And it's also about closure. I've said about all I have to say. The fact is, I use climate and past events to formulate a base for my forecasts. I understand how modeling can go astray, and I understand how knowing what happened before raises questions about what is happening now. But what else can I say? This book is a means for that.

Above all, it's for the love of the weather.

It comes down to this: I guess I'm just a prisoner of the weather, because it captured my heart from the day I was born.

Table of Contents

Chapter 1

So What Do I Believe About the Climate?

Three articles I have written summarize my position. The first I want to highlight is "The Grand Slam of Climate" in which I added a fourth dimension to my original theory: "The Triple Crown of (Climate) Cooling."

From "The Grand Slam of Climate," written in January 2015:

> I introduced something on "The O'Reilly Factor" several years ago called The Triple Crown of Cooling. I called it that because back in 2007 I thought a period of cooling lasting 20 to 30 years would start, resulting in global temperatures returning to 1978 levels by 2030. I also introduced the concept that this cooling may cause a "time of climatic hardship" — in other words, the natural process of cooling after a process of natural warming could produce an uptick in extreme events. The increase is not clear, though one can argue it is occurring off the East Coast. The Atlantic remains in its warm cycle and will continue to be for several more years, so the coastal water is warm. It is the reason I am very worried about the East Coast with hurricanes similar in magnitude to the storms of the 1950s, though it has not yet occurred. That's right — Irene, Sandy and Arthur can't hold a candle to eight major hurricane hits in seven years. None of the storms was major.

The fact is, winters have been getting colder in the U.S., as data compiled by NOAA's National Climatic Data Center shows. And it's this onslaught of *colder* temperatures that is likely the cause for any uptick in snowfall near the East Coast. Once the Western Atlantic cools again, the snows will go back toward normal.

Therefore, one should look for natural clashes near the East Coast — the key word here being *natural* — as temperatures have started falling over the past 10 years.

[Note: This article was written in 2015. Temperatures have recently begun to drop, but we want to stay true to the time this was written.]

The point is that all this was introduced years ago during a time when the missive was that winters won't be cold and snowy and the ice cap is melting away. Now I will make another forecast in a five-year increment: At least three of the next five winters will be warmer than average across the eastern U.S. The Great Plains will oscillate, and the core of the coldest winters will be in the western U.S. Let's see how I do. **[Already two out of two winters have been warmer than average in the eastern U.S.]**

Even though severe cold and enhanced storminess will rule the roost over the next couple of weeks — and we think spring is going to be very late this year for much of the country — spring training for baseball is around the corner, so I decided to rename my "climate for 'dummies' idea" The Grand Slam of Climate.

Let's ask these questions:

1) Does the sun have a far greater effect on the climate than CO_2?

2) Do the cycles in the ocean, with the vast amount of the earth's heat stored in them, have a far greater effect on the climate than CO_2?

3) Do stochastic events (ex-volcanoes, etc.) have a far greater effect on the climate than CO_2?

And now I have added a fourth leg, The Grand Slam:

4) Does the very design of the system have a far greater effect on the climate than CO_2?

Quantifying CO_2's effect, with its increase of only one molecule out of every 10,000 molecules of air over a 100-year period, against The Grand Slam of Climate, especially in light of the earth's having had ice ages at 7,000 ppm and warmer times at 250 ppm, is grasping at straws at best. Then again, desperate people zealous about another issue would do that too if they felt this would help them get their way.

Just ask yourself the questions above and see what you come up with. It's not that you're dumb, it's just that climate alarmists think you are. So let's humor them a bit.

By the way, here's a fun thing to think about: Mars has an atmosphere with the same percentage of CO_2 as Venus, but it's much less dense. So why is Mars so much colder than Venus? And just why do those Martian icecaps shrink for years only to expand again? These questions are out of this world; the ones in The Grand Slam of Climate are not.

My piece "A Short Summation of My Climate Position" adopts a very balanced idea. It was written on May 23, 2016:

> Earth is warmer now than in the late 1970s, the start of the satellite era.
>
> This can be explained largely by the turn of the Pacific and the Atlantic to their warm cycles. This is the "bathroom shower theory" that I have used many times. Turn on a hot shower, and the bathroom will heat up until an equilibrium is reached. When the Pacific warmed, and the Atlantic followed, we came off a period in which they were cold in tandem. It is perfectly logical that with the oceans and especially the tropics — the thermostat for the globe — warming, the air must warm until it reaches an equilibrium, which it appears overall to have done until the last 20 years, when it did not warm.
>
> The recent El Niño temperature spike has already started to descend. Given that cold, dry air is easier to warm than already warm, moist air, the natural place the warming shows up most is where the atmosphere is driest and coldest. The biggest warming has been in the Arctic during winter. Recent summers in the region have actually been a bit below normal. There is no argument here. The question is how

much of this is because of the increase of greenhouse gases, specifically CO_2.

Another question: Will temperature measurements return to the levels they were in the late '70s? That question cannot be answered until the PDO/AMO shift back in the coming years and we can observe what happens when we go through the entire cycle. Even then, the state of the oceans today is a product of centuries of back and forth, making it very difficult if not impossible to assign a specific value to CO_2's role or blame it for any event, short or long term.

No one denies the climate changes. And *I would challenge* anyone who denies the climate changes. Of course it does — swings have occurred since well before man appeared on the planet.

In fact, it's quite evident that not only does the climate change naturally but the warmer it is, the better. Earlier warm periods, which dwarf today's warmth, were climate *optimums*. How is it that previous warmer times were referred to as climate optimums? Let's look at the definition of optimum.

Used as an adjective, optimum, according to Dictionary.com, means "most favorable or desirable; best."

As a noun, it means "the best or most favorable point, degree, amount, etc., as of temperature, light, and moisture for the growth or reproduction of an organism."

Will the term "optimum" need to be adjusted, or will the temperature need to be adjusted down to fit the current missive of impending disaster?

The "climate change denier" label is a straw man argument that is designed to isolate, demonize and destroy people with false labels.

The whole argument as to what is best for us going forward is simple.

1) How much is man responsible for variances that were previously exclusively natural?

In my opinion, most of the warmth today is likely natural given the tiny amounts of CO_2 relative to the entire system, of which the oceans have 1,000 times the heat capacity and are the great thermostat of the planet, taking centuries of action and reaction to reach where they are today.

2) Is this worth the draconian reactions that will handcuff the greatest experiment in freedom and prosperity in history, the United States of America?

3) This question may arise: Would not the cost of adaption rather than trying to preclude an ordinary recurrence be a sounder fiscal response?

Remember, Barack Obama's EPA administrator said all these measures would save .01°C over 30 years and that it was mostly intended to be an example for the rest of the world. Color me skeptical that the rest of the world is going to follow. Instead, it will take advantage of repercussions on the American way of life. Not every nation is our friend, after all, if you examine the real world. No one is against any form of clean, safe, cheap energy. I am against economic suicide like we have seen in Europe, which will take economic peace and prosperity from future generations.

Finally, the last blog in this series contains five devastating graphics that blow away my opponents' whole missive. From "Shifting Gears on My Climate Debate Contributions," published on June 19, 2017:

> There are two blogs I have written that really describe my views on climate.
>
> The first is "The Grand Slam of Climate." In this I explain why the big macro drivers that have always governed the planet's climate are far more likely to still be doing so today.
>
> The second, "A Short Summation of My Climate Position," lays out three points that sum up for me this whole issue.
>
> I see nothing extreme or irrational about either position. Nothing I would take back now. It's simply looking at the past and arguing that, given the entire planetary history and reasons for what has happened so far, it's highly unlikely CO_2 is now the control knob of the climate.
>
> But I want this to be a third contribution to that group. From now on, I will focus on how weather events, when they are brought up as "the worst ever," have happened before. I am particularly nervous about this hurricane season, as sea surface temperatures off the East

Coast are very close to the period 1954-1960, when eight major hurricanes ran the coast over seven years! Given the media today, if a repeat occurs, we will need someone who knows where to go to find examples, and that is my niche.

But being obsessed with fighting an issue that to me has wasted far too much of our nation's time and treasure is no longer something I need to do. It is now akin to the "Monty Python" scene where the Black Knight is guarding the bridge. How many times can you show and refute something before it's simply getting in the way of what you are meant to be doing?

No matter how many times they're disproved, climate alarmists keep trying to hold a ground they don't have.

There are two serious items I wish to share with you.

The first, "To Put America First Is to Put Our Planet's Climate First" — authored by Istvan Marko, J. Scott Armstrong, William M. Briggs, Kesten Green, Hermann Harde, David R. Legates, Christopher Monckton of Brenchley, and Willie Soon — is a tour de force in debunking the climate change hysteria that even people behind it have admitted is intended to destroy capitalism. Nothing in the article summed it up better than this:

"Ms. Christiana Figueres, executive secretary of the U.N. Framework Convention on Climate Change until [2016], openly stated in 2015 that the goal was to overturn capitalism — in her words, 'to change the economic development model that has been reigning for at least 150 years, since the industrial revolution.'"

Which gets me to the second piece. Ms. Figueres is not alone. It's just that climate alarmists are getting bolder about it.

There are people driving this who are now revealing their true motive. What does that say about this debate with which we exhaust ourselves? We are debating a point that, no matter what the answer is, to the side driving it, it doesn't matter.

Back to the point of how to make more progress for mankind. Five minutes of Alex Epstein's April 2016 testimony before Congress was breathtaking to me in terms of making the moral case for fossil fuels as the way to advance the human condition.

Just knowing what is stated in both these presentations should be, at the very least, a cause for skepticism on the whole man-made global warming theory. And remember: When I say skepticism, I mean questioning. If you're not open-minded enough to see the rational basis for questioning, then you have to *deny* the scientific methods that have led to mankind's advancement: questioning, searching and discovering. Without that, there is no *true* progress. Keep in mind, this is against a backdrop of stated goals that basically use all of this to achieve an end; the means are intended to deceive to get to the final point.

In the entire geological history of the planet, there has been no known linkage between CO_2 and temperatures. I show this in the section on illustration along with the fact that warmer times were climate optimums.

Mankind does better overall when it's warm. Most of the current warming has been in polar regions during wintertime. Where most people live, the warming is much smaller, and in the tropics, it's negligible. In fact, Dr. Richard Lindzen points out what I have been saying for several years: This may lead to a decrease in extreme weather, because the temperature difference north to south (south to north in the Southern Hemisphere) means there is less "zonal potential energy" (roughly put, temperature gradient) to fuel storms. In the case of hurricanes, the decrease in global ACE may be because of changes in the global wind oscillation from changes in surface pressures by disproportionate warming. That is certainly different from Al Gore's prophecies of doom and gloom. Furthermore, it shows that he was not thinking about what he was saying. Or maybe he was simply doing as he was told. In any case, the proof is there. A forecast was made, and it busted, horribly.

When it comes to catastrophic sea level rises, Greenland is the supposed canary in the coal mine. It is not a frozen ocean like the Arctic, rather it's composed of land with tremendous amounts of snow and ice on it. Using ice core samples as a proxy for temperatures, just like tree rings from a regionalized area can be used, we see how Greenland saw far greater warming during olden times. Over the past 10,000 years, the current "hockey stick" period certainly shows up. But compared to other previous warm periods, it is dwarfed. So why is this metric discounted in lieu of a single tree ring study? Wait a minute — *there is not just one tree ring study!* Liu's tree ring study, which originated in China, shows that this period is no more spectacular than previous ones.

The warming is certainly there, but so are other warmer times that are not in the infamous Hockey Stick study. Yet why is one tree ring study cited religiously while another, less unnerving one is ignored? Perhaps it has to do with Ms. Christiana Figueres' telling remark above.

Finally, to Mr Epstein's point. Global progress as measured by per capita income and life expectancy has taken off in the fossil fuel era. Coupled with the increase in population, it means more people are living longer and easier lives than before fossil fuels. Is that not a good thing?

Science is being used as a means to an end. There are wild fights over weather patterns and storms each day, but the patterns and the storms supply answers. Large-scale macro events have far more weight than what is overall a micro contributor. But the battle between economic systems, in spite of my obvious bias toward capitalism and competition, is something others should be fighting for tooth and nail if they believe in it.

The conclusion is that the climate has been used as a tool for something that is not related to science. It is very valuable and underappreciated by itself! But it does not have to be a means to an end to have the great value it does. And perhaps that's another problem — if you enhance the value of something that is your lifeline, then you enhance yourself. The ultimate irony? The system we have now, which has given people the ability to spread their wings, is the system climate alarmists are now trying to destroy and replace.

These three blogs form the core of my beliefs on the reasons for today's global warming. It comes down to this: For you to believe the anthropogenic global warming argument, you have to believe that the increase of one molecule of CO_2 out of every 10,000 molecules of air over a *100-year period* is now controlling the climate for the first time in known history. It's controlling the sun, the oceans, stochastic events and the very design of the system that encourages conflict and resolution. Climate alarmists are setting abstract goals without ever defining what the extreme long-term average temperature of the earth is (no one knows), what the perfect temperature of the planet is, or what the perfect level of CO_2 is for life on the planet is. Which is the reasonable position — mine, or the one that defines no boundaries and asks for blind faith to follow?

Finally, as anyone who is truly seeking the truth knows, it's important to look at the counterarguments. One of two things should happen when someone argues with you on an objective playing field: 1) you find out you are wrong and change, leaving you better off, or 2) you find out you are right, leaving you stronger. But I ask you, have you ever seen any of the major proponents of the "CO_2 is the climate control knob" theory do a public self-examination of why they might be wrong? Of course not. Because when their very livelihoods are on the line, there can be no being wrong. However, I must deal with the facts at hand no matter the outcome. I highlight this contradiction in "Is There Anything in the Global Warming Debate That Would Convince Me I'm Wrong?" which I wrote on April 4, 2015. I do question and do look at doubt and challenge as tools to get the right answer. There should be skepticism! Therefore, I question myself. From the article:

There is a constant process I go through, whether it be forecasting or training with weights. Unless challenged, you do not improve. The challenge is listed in the above title. I must always seek the right answer. I'm not getting soft, I'm simply using this methodology to either confirm or deny my idea.

So, is there anything in the global warming debate that would convince me I'm wrong?

As a matter of fact, yes.

The PDO, which is tracking nicely with the 1950s, turned cold in the early and middle part of the decade, warmed late (as it is now) and then turned cold again. The AMO, which is in its warm endgame now, turned cold, but temperatures did not fall. However, since the 1997 Super El Niño and then the flip of the PDO, there has been no increase in global temperatures overall in the last 20 years.

That's one factor.

Another is the top of the stratosphere, which has been cooling since the late 1970s. But since the Pacific flipped, it's warming! That to me says there was an expansion in levels below it (warming). Now the question becomes: Is this warming driven by the cycles of the ocean (which I believe) that reverses as it cools, leading to stratospheric warming? (We may be starting to see that now and the result would be stronger cold invasions over the continent.) Or is it truly because of the increase in human-caused greenhouse gases? (There's nothing we can do about nature, though one could argue that CO_2 is entirely natural since it comes from plant and animal life anyway.) From 1979 to 2008, it cooled over the Arctic region. But over the last several winters, it is reversing and warming.

I am watching the Southern Hemisphere because ice expansion in an area *surrounded* by water seems to be a perfect counterbalance. Total global sea ice, since the decadol shift in the Pacific around 2007, has more or less returned to normal!

Finally, I am waiting for someone, somewhere, to *quantify* global water vapor, which I believe is the true measure of the climate system. Over the tropics, where hot spots are supposed to occur, water vapor varies directly with the ups and downs of the tropical oceans. Understanding this process and its relationship to global cloudiness would be huge.

Since the PDO flip, it has dried out in the middle and upper levels of the tropics, and this is in direct opposition to the trapping hot spot theory.

Objective satellite-era-based measurements in conjunction with oceanic cycles would be a much better way of measuring where the climate is going. Obviously, a sustained increase in water vapor, by far the most prominent greenhouse gas, would be indicative of a change that is meaningful. The bulk of the "warming" has been where it's dry and cold. In the long-running register of record highs, it's not getting hotter. Additionally, the current hysteria that everything is the worst ever is a function of several non-scientific, highly subjective variables. Chief among them is the fact that we can observe almost everything now, and people are not acquainted with what has happened before in detail. If we want to continue to follow along with global temperatures, then it's a simple test: Watch what they do over the next 15 to 20 years as the oceans cool.

These are things I look at that can lead me to say, "Joe, you are wrong."

The test period is here.

For me, this debate is far from settled. And it's the atmosphere that should settle it, not agendas.

I ask people who don't see things my way: Is there anything that can challenge your position on this? If not, then your position is dogma — very different from what is needed to strive for the correct idea on this matter.

We must constantly question our viewpoints. I have strongly held beliefs on this issue, but I do question myself. The only way to advance is to meet challenge and doubt with objective analysis that does not depend on an agenda but rather the search for the correct answer, no matter where it leads.

Chapter 2

Five Illustrations That Question, if Not Refute, the CO$_2$ Missive

There are many charts that refute the "CO$_2$ is the control knob of the climate" argument and the whole missive that the planet is heading for doomsday. Quite the contrary. The following five charts show me the opposite.

The first is a geological chart dating back millions of years. It shows no linkage between temperatures and CO$_2$.

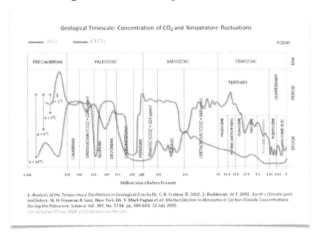

I often tell people that this is like a 3D book. You have to squint your eyes to see the picture. In truth, it's fairly straightforward — there is no linkage, and the assumption that the upturn today in CO$_2$ is causing the current warming is akin to saying that if one sees more squirrels in Central Park, it means it will be a snowier winter. There is simply no basis in fact when looking at the past.

Past warmer periods were called climate optimums.

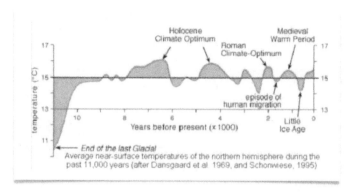

Average near-surface temperatures of the northern hemisphere during the past 11,000 years (after Dansgaard et al 1969, and Schonwese, 1995)

As stated in the previous chapter, optimum means "most favorable or desirable; best."

Therefore, climate optimums mean that warmer times have been most conducive to life on the planet. Has anyone noticed that very few people today live in the polar regions?

The global warming Hockey Stick, of course, was a big deal. The tree ring study from which the Hockey Stick was derived had one problem — it wasn't entirely made up of tree rings. It was up until the point that the tree rings no longer showed warming, when in fact the global temperature continued to rise. That part was replaced with instrumentation. The problem, of course, is that if the tree rings no longer showed warming when it was continuing to warm, that means other periods could have been warmer than currently shown. Hence the controversy.To add to this, there are other tree ring studies which show that the current period we are in is not that big a deal — no warmer than other times. One example: Liu's study in China.

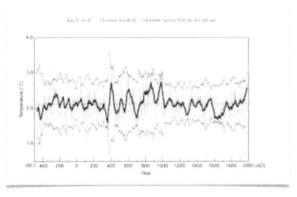

Scientists attempt to recreate temperatures by using tools like tree rings. When you study the weather, you understand that long-term patterns, even when they're on the order of a few weeks, are linked. We call these teleconnections.

For example: If it's warm in one place, it may be cold in another. In climate, when long periods of warmth or cold show up (on the order of decades or centuries), it means the earth is in a warm or cold spell. Ice core samples from Greenland are also a proxy.

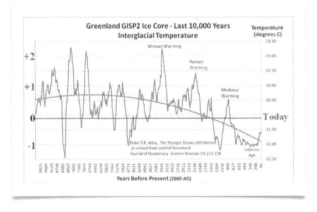

Greenland, when it's warm and in a period of melting, has been used as the impending canary in the coal mine. But a look at samples shows that the current warmth is nothing remarkable and is in fact timid relative to other warm periods.

Here is the critical question: Why is the public and media exposed only to information stacked on one side of the argument when the evidence that questions, if not refutes, the whole argument is there for all to see?

Finally, the true hockey sticks of the fossil fuel era have to do with human progress. Population, life expectancy and personal wealth have exploded since fossil fuels became the prime source of energy on the planet.

In the scientific method, something must be true all the time for it to be factual. Example: We know there is gravity because objects fall to the earth. We know water freezes at 32°F. This is a fact. Therefore, the argument that CO_2 is the climate control knob and the planet is heading toward some atmospheric apocalypse is easily debunked by these charts. There are many, many more, but to avoid bogging down the book, we'll stick with these five.

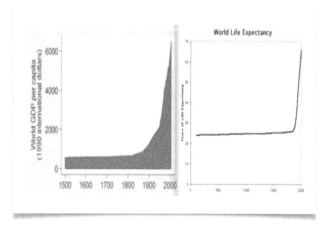

The first four charts verify the idea that the sun, the oceans, stochastic events and the very design of the entire system far outweigh the increase of one molecule of CO_2 out of every

10,000 molecules of air over a 100-year period when it comes to what controls the climate. The last chart verifies the fact that man is better off because of it. Finally, the conclusion is almost scriptural: There is nothing new under the sun, including men who wish to create much ado about nothing. Yet we are being pushed to believe disaster is imminent.

Remember, as H. L. Mencken said: "The whole aim of practical politics is to keep the populace alarmed — and hence clamorous to be led to safety — by menacing it with an endless series of hobgoblins, all of them imaginary." And, "The urge to save humanity is almost always only a false face for the urge to rule it."

Chapter 3

The Weaponization of the Weather

While I think it is absurd to label what should be a debate among people interested in the global climate as a war, the fact is that to people who are using climate as a means to an end, this is a war. Part of the reason is that climate alarmists have no actual conception of what real war is — where people, many of them innocent, fight and die. Yet they fancy themselves as some kind of hero in a war. This hit me when I read Dr. Michael Mann's book, *The Hockey Stick and the Climate Wars: Dispatches from the Front Lines.*

I read everything Dr. Mann writes as I consider things on an issue-by-issue basis. The Hockey Stick made him famous, but he did a lot of great work beforehand. It's just that most of it was not something that could be used for an agenda. That's right — there is a lot of great work he has done. I disagree with him on this matter and the portrayal of it as a "war." That he or other people are demonized because of their climate stance is common today, as the totality of a person's journey means little anymore. However, the problem is that if you let your work on an issue become your life, then rational balance is taken out. What you own owns you. Therefore, an attack on your work is an attack on your life. Consequently, the people pushing the climate agenda, many of them with a socialist bent, do believe they are under attack both personally and as a group, so it's a war footing.

Many on my side engage in attacks. This takes out the rational balance needed for a scientific debate. It's personal. In my case, each forecast does not define my life's work, which is the weather. You can't take the weather away; it's there every day. If you say, "Joe, you busted on this forecast," it's not like I don't have a chance tomorrow to show what I can do. If the Hockey Stick is proven wrong, then what becomes of the people supporting it? If scientific debate becomes a matter of life and death and you view it as a war, I suggest you spend a year actually fighting in a bona fide war. You would look at it very differently.

The climate agenda comes with a deep-seeded desire to be a hero. All of us have it, really. But we don't engineer our circumstances. In true battle, your reaction is a measure of where your character is at the time. What is going on in the climate debate is that a whole group of people who want to be heroes are engineering what I think is a phony war, and part of it is because they can be the so-called heroes of that war. It's a debate — nothing more, nothing less — and when you look at it that way, you can be objective. The problem is we are not dealing with people who look at it this way. It's life and death to them. They are at war, and they are weaponizing the weather!

The very title of his book makes it clear that Dr. Mann views this as a war, as do most of his comrades, and this is a heroic fight against all odds, when actually the bulk of the money and media support is on the side of those pushing man-made climate change. This should not be a war. But when political, economic and social ends are at stake, the means to achieve them involve the kind of soft tyranny (and lately threats of even greater tyranny — trials for people with whom climate alarmists disagree) that are despotic in nature; things people have resisted and gone to war over.

Again, as H. L. Mencken said: "The urge to save humanity is almost always only a false face for the urge to rule it."

While people on my side wish for reasoned debate in the normal scientific process, we have people on another side who want no part of that but instead seek to wipe out all forms of debate. As such, they have weaponized weather by using every meteorological example they can find to bolster their point, knowing full well an uneducated public and a "history began 10 minutes ago" society is quick to run to the last thing they see.

Here is what I have written to demonstrate this, beginning with "Hurricanes and Climate Change," published on August 31, 2012:

> Hurricanes have become a focal point in the climate change debate. The assertion that every problematic storm is a sign of global warming is absurd. However, the counterargument that storms are less than what they actually are is incorrect as well. While it seems to be a knee-jerk reaction to the falsehoods generated by AGW propagandists, it is not helpful to the debate.
>
> Hurricane Irene was a case in point. The Left used the storm as a pin-up for global warming, but the Left is also most likely unaware of what occurred in the 1950s. The pattern then, with a cold Pacific and warm Atlantic, has been a mainstay of my overall pattern ideas since 2007. Eight *major* hurricanes struck the East Coast in the 1950s from the Carolinas and points north.
>
> The bottom line: Storms like Irene are to be expected in this cycle, yet AGW propagandists won't acknowledge this fact. The knee-jerk reaction to counter their statements is to try to downplay Irene, calling it a strong nor'easter. This counterargument is also incorrect and not helpful to our cause. It is important to fight back with the truth. There are many nor'easters every year, and none of them has become the seventh-costliest hurricane in U.S. history.

Nor'easters don't cause $15 billion in damage. Nor'easters don't knock out power to five million people. Part of the reason for the expense of Irene is because more people are now living in harm's way. As a result, hurricanes that make landfall in the present will cause more damage than hurricanes that made landfall in the 1950s. It has nothing to do with storms being stronger.

Even with Irene and a warm Atlantic, the ACE index for 2011 was significantly lower than previous years. Despite the fact that the National Hurricane Center names more storms, which skews the frequency upwards, the ACE is still low. The fact of the matter is that hurricane frequency and strength are a result of natural variability, not global warming.

When the great hurricane of 1938 drove 15 feet of water into Providence, nobody was shouting that global warming was the cause.

In summary, there will be a lot of disinformation on hurricanes and climate change with each storm, but more so in the coming years of a naturally enhanced cycle. The facts are clear: What naturally happened before under similar conditions can and will happen again and should be expected. This has nothing to do with man-made global warming and everything to do with nature.

Hurricanes and tropical cyclones are a natural weapon when you deal with a populace and media that have no idea about the past. History combats climate hysteria if one is educated in it, but climate alarmists understand that for most of the population, they have a free pass to blow anything they want out of proportion.

Hurricane Sandy was seen well in advance. As summed up by Sean Hannity at the time:

"Joe Bastardi … warned us about Hurricane Sandy nine days before landfall. [His team] provided unwavering forecasts for a landfalling hurricane around NYC, and we were the first to know that this unprecedented event would occur."

But like all these events, you know what's coming after. The fact is that five days beforehand I was being ridiculed for the forecast, and two days prior Mayor Michael Bloomberg said that New York City would be open for business during the storm. In 2003, after Isabel hit the North Carolina coast, I started doing talks on the "Sandy Scenario." It could have been worse if the storm was farther south, driving water up Delaware Bay while flooding rains were occurring in the Delaware River Valley. The intersection of the storm surge and the flooding could have met over the port of Philadelphia.

On Oct. 25, 2012, I wrote about Sandy, which, by the way, at 5 a.m. that morning was officially forecasted to be 300 miles out to sea as a post-tropical storm. From "This Storm Is Long Overdue!"

> Since the weekend [Oct. 21] I have been loudly warning that a historic storm is in the making. For years I have wondered why a storm like what I have forecast — a hurricane hit on the mid or north Atlantic coast from the southeast — has not occurred. In this case a tropical cyclone is entrained into a massive early season cold air mass and is forced to turn northwest into the Mid-Atlantic states. In fact, in talks I give about hurricanes, I show the example of this kind of storm, which I call the "Philadelphia Story" — a hurricane hits the Mid-Atlantic states from the southeast near or south of the mouth of the Delaware Bay. It

sends its storm surge up the funnel-shaped Delaware Bay — much like the 1938 hurricane did in Providence — and that rising storm surge is met by flooding coming down the Delaware River. The flooding is worse than it would have been because the lakes that were not dammed up years ago are now, so they must release their water downstream into the river. Additionally, such a storm would devastate the Jersey Shore worse than in 1944.

But here is the key point: *It is a product of past individual events we have seen combining in one place.* Instead of asking why something like this is happening, I have always wondered why it has *not*!

I've been saying since the start of the hurricane season that the end of the season this year would feature a tropical cyclone coming out of the Caribbean that would threaten the East Coast. Just like the season started with in-close storms, we felt it would end with them, too. Three weeks ago, we again told clients the pattern was setting up for this.

On Sunday we put out a chart hitting New York City from the south-southeast with this storm, even before it was named, with a 75-knot hurricane early next week. My forecast today has been adjusted a bit south but calls for this to go into New Jersey with a major snowstorm on the western side from western Pennsylvania into West Virginia. Clients have been told that this is a multibillion-dollar storm in the making where it hits and that this could be the worst historically in some places along the Mid-Atlantic coast. While there is still room for me to be wrong, the window is closing on the chance of this escaping. And what I have always worried about looks to finally happen.

It will not be global warming but the combination of natural events that will cause it. In fact, my father, a degreed meteorologist, would show me tracks from hurricanes when I was a kid and say that all one needs to do is move that track a couple hundred miles and what appears to be ready to happen now would happen. The point is that this was seen as something that could happen years ago, even by people in the field before me. If it does occur, it will be because of nature, *not* man.

That this warning shot — and I issue Climatic Ambulance Chaser watches all the time — is ignored by the Left shows me that it really doesn't care. And since it has the press on its side, it knows it won't get called out. How is it the guy who doesn't believe that CO_2 is the control knob of the climate forecasted the very thing leftists tried to exploit once it happened?

As a side note, Sandy may have been one of those events that changed the course of history. Whatever momentum presidential candidate Mitt Romney had came to a halt for several days, and the photo ops of the sitting president and the governor of New Jersey were a big deal. In fact, a few months earlier I wrote in "Weather as a Number One Factor" how John F. Kennedy's assassination and 9/11 were aided by great weather, and both events may have changed history.

Weather is the number one factor in everyone's life. Think about it. No matter where you live, or what you do, you have to deal with the weather every day. Perhaps that is why the irrational Tower of Babel belief has evolved into the idea that somehow mortals on Earth can control what the heavens do. The events of Nov. 22, 1963, and Sept. 11, 2001, are two where the weather played a role and changed history.

Consider this: If the rain that was forecasted to hold on all day in Dallas the day President Kennedy was to speak there had continued, the motorcade would never have had the top down for the president to be exposed to the assassin's bullets. Who knows what the rest of the Kennedy administration would have brought. As a fan of John Kennedy who is well aware that he is a man with his shortcomings, perhaps there would have been a very different path as far as Vietnam went. Perhaps there would have been no Nixon presidency. The faith people had in our nation was trashed in the '60s and then with Watergate, and I think we are still paying the price today. But if it was raining, who knows what would have happened, right?

But it wasn't raining — the sun was out, and the assassin's bullets hit their mark.

On 9/11 the sun was shining brightly as high pressure along the East Coast and strong sinking air outside of major Hurricane Erin well offshore allowed for a spectacular day on the East Coast. If Erin were closer, chances are the weather would have been less hospitable or perhaps there would have been flight delays. As it was, the storm was in the perfect spot to allow the usual sinking air to create ideal conditions for the hijackers to do what they did. The course of history changed, forcing the sitting president to wake up every morning for the rest of his presidency thinking about how to prevent another 9/11. The hit our economy took was staggering, and we are still paying for it in many ways, some clear (air travel), some not so clear. But again, the weather played a big role, and the nation reacted in a way that a major part of the population, after an initial outpouring of support, lost faith in those trying to lead.

Erin was out at sea, and the hijackers hit their mark.

Because I love the weather I may be guilty of overstating these two events. But there is nothing else I can think of in my lifetime where the weather being different may have had such a drastic effect on the course of history. We will never know what a Kennedy presidency and re-election would have led to in the years following, just like we will never know what a George W. Bush presidency would have been like without the burden of 9/11 constantly hanging over his head and a press that considered him stupid and illegitimate. But one thing we do know — nature, not man, has the final say in the course of events where nature is involved.

From Robert Burns:

The best laid schemes of mice and men
Go often awry,
And leave us nothing but grief and pain,
For promised joy!

Sandy became the third.

Droughts are a weapon that climate alarmists love to distort, so I wrote "Widespread Dryness: Been There, Done That" on Jan. 31, 2013:

I have been on national television several times over the past six years saying we would go into a pattern similar to the 1950s with drought and heat centered over the nation's midsection with enhanced hurricane threats on the East Coast. The reason is simple: We have the same type of natural cyclical pattern we had then, with the Pacific going into its cold stage (cold PDO) while the Atlantic is in its warm stage (warm AMO). The warm AMO is also responsible for the Arctic ice cap being smaller than it was when the AMO was cold. But debunking one climate change myth at a time is all I am going to do here, so let's take a look at the well-forecasted drought that climate alarmists are playing up.

If we look at the years when the Pacific went into its cold stage while the Atlantic was warm (cold PDO, warm AMO from 1951-1960), we find the nation's midsection is quite dry, as it is now.

Conversely, when the opposite occurs, the U.S. is wet!

Texas was wet was when the global temperature was climbing, exactly opposite of what President Barack Obama said last year during the Texas drought, when the earth was cooling. (Has he called the governor of Texas yet to say he misspoke? Don't hold your breath.)

When the Pacific is in its warm cycle, as it was from 1978-2007, El Niños are more frequent, so the input of tropical moisture into the U.S. allows for more rain. And guess what happens when you warm the tropical Pacific after it was in its cold cycle? Since the oceans have a thousand times the heat capacity of air, the air warms when the oceans warm. It's simple climate cycle theory.

As soon as the earth warms up enough so a balance is reached, it levels off and then starts to cool (remember, we started the satellite era at the end of the last cold cycle in the late 1970s, so a warm PDO can be expected to warm the atmosphere). The cooling starting now will probably take temperatures back to the levels they were in 1978 by 2030, once the Atlantic shifts to cold in five to 10 years.

It was drier in the 1950s than it has been now, suggesting the worst is yet to come! I also suggest the dryness is being exaggerated relative to what has actually happened before. Remember: Drought breeds heat, since dry ground means the sun does not have to waste energy evaporating water out of the ground. The ground heats faster, so the air above it does too. **[Like the 1950s, those droughts broke, even in the face of climate alarmists yelling there was a perma-drought starting.]**

It's not rocket science, and it's certainly not CO_2. Yet why does the mainstream media simply swallow the extreme weather idea without even looking at the naturally occurring factors that are obvious?

Remember: The hurricane aspect is there. I have said many time since before 2010 that we are entering a perilous time on the East Coast.

Climate alarmists are trying to rely on the fact that most people don't know the meteorological history of the planet and that we now can observe everything to drive their agenda. That has precious little to do with actual protection of our environment. Instead, it's a control issue. As of now, there is no law against arming yourself with knowledge to defend against this, and my suggestion is that you continue to do so. Not doing so will limit even more of your freedom.

What's astounding to me is that climate alarmists don't get called out on these things. The so-called perma-drought earlier in the decade reversed similar to the way it did in the 1950s. Even though I forecasted that to happen, no one said anything about the bogus forecast they were pushing. (Dr. Judith Curry refers to them as the Climatariat. They're like a parasite in that the amount of money it takes to sustain them is all take and no giving back to the host, who happens to be the American taxpayer.)

One of the biggest weapons of mass deception being used is the anything-that-happens-is-because-of-what-you-said device. The following piece, "Is Global Warming Causing More Snow and Less Snow at the Same Time?" was written in March 2013 in response to the fact it was snowing an awful lot, and climate alarmists were noticing. They were making a big deal about how warm it was the previous March, claiming that also was a sign of global warming.

In the Woody Allen comedy classic "Bananas," there is a scene where the CIA is sending U.S. troops to fight on both sides of a revolution because they're afraid of being on the wrong side.

While many of us laughed at such things then because of the absurdity of it, we find that the same kind of strategy is being used by climate propagandists who seek to claim every idea they have is the correct one.

In this piece I will examine a recent claim that climate change is causing extremes that rational people know is not a sign it's getting warmer (such as bigger *snow*storms — snow is a sign it's cold). Irrational people are the ones who will claim the opposite answer of what they said before was right, because they will use any answer to justify their ideas.

In a second piece later on I will explain how cooling in the tropical Pacific leads to drought and heat in the U.S. The downturn in global temperatures that has started with the cooling Pacific means it cools globally, but the initial result, like in the 1950s, is a warmer U.S. — opposite of the alarmist spin.

Keep in mind, their argument when it's *not* snowing is that it's getting warmer. But their argument when it *does* snow is also that it's getting warmer. So no matter what happens, climate alarmists blame a warming climate. Wouldn't you love to have a class where no matter what answer you put down on a test, you get to say it's correct?

Meteorologist Joe D'Aleo put it this way:

"Snow has a negative feedback on temperatures. It cools the air, lowers the thicknesses/heights, and often shifts the jet stream and storm track so as to favor more snow."

Extrapolating that logic on more snow, one can see that if ice makes it into the Gulf of Mexico one of these winters, it will also be because it's getting warmer.

But here is an inconvenient truth: *It's not getting warmer.* The satellite data proves it.

With temperatures having leveled off, and even getting colder since the statements were made about it being warmer with less snow, the idea that warming is causing more snow is plainly wrong.

For the record, here in the U.S., winters have been getting colder.

Now, do you see me screaming about it getting colder, resulting in more snow? I could, but I don't. But it's certainly not getting warmer in the U.S. Data from the National Climatic Data Center plainly shows that winters have cooled across all of the U.S. over the last decade. The global temperature has leveled off or declined over the past several years. So if it's snowing more, it can't be because it's getting warmer — because it's not.

There is a natural cap on what temperatures can do, and none of it has anything to do with man-made climate change. Man can only do what nature allows him to do.

The fact that my opponents said what they did a few years ago and when the opposite happens they try to use that as evidence speaks volumes as to their grasp of "the science." Who was it that said, "There is a sucker born every minute"?

> Perhaps that is who they are appealing to with their message, which, as near as I can tell, is: "No matter what happens, it's because of what we say it is."

And that is their secret: a weapon of mass deception, with the media as the willing supply line.

The year 2013 was a bad one for climate alarmists. The spring was cold, and they had to resort to using the argument that too much snow and cold was a sign of climate change. The summer was not hot in the nation's midsection as the previous three had been, and the "perma drought" for the Southern Plains was starting to reverse. But never fear — if you wait long enough, somewhere on the planet will have an event you can claim as your very own. And so when Typhoon Haiyan appeared, the schemers were ready.

Never mind it had been a non-season in the Atlantic and the Western Pacific was below normal. The one big storm that did show up was used relentlessly as a propaganda tool.

On Nov. 14, 2013, I wrote "The Shameful Deceit on Haiyan."

> The people claiming that climate change is the cause for what has been labeled by the meteorological centers as a *Category 4 landfall of Typhoon Haiyan* in the Philippines are either ignorant of the facts or lying about them. Earth to climate alarmists: What you see is what Category 4 storms do to buildings that can't withstand winds close to 150 m.p.h.

> What's more remarkable: a major typhoon hit on the Philippines — a place that's been struck hundreds of times by typhoons, dozens of which were major — or hits by major Category 4 hurricanes Flora, Cleo and Inez in a four-year

stretch between 1963 and 1966 on Haiti? And why don't we throw in Betsy, a strong Cat. 4 at its peak that struck the Florida Keys and then the central Gulf Coast in 1965. That's four storms in four years in a geographical area that's less than the size of the Philippines (Haiti to Florida) *in the Atlantic Basin, which only has one-third the activity of the Pacific on average!* What's more remarkable: the weakened Haiyan going into North Vietnam as a tropical storm (why didn't it come roaring back?), or the typhoon that hit Haiphong in 1881 and *killed* 250,000? Or the Bay of Bengal, where multiple cyclones have killed over 100,000 people several times, all when CO_2 was well below where it is now?

For those who want to see either the ignorance or lying on display, look at Flora, Cleo, Inez and Betsy and the destruction they caused.

By the way, in the spirit of Fidel Castro — who blamed the U.S. for stalling Flora over his island to destroy him — using weather for political motives is nothing new with people who care little about the truth. It's just ramped up to levels that would make Castro look like a novice.

If you're aware of this — three major hurricanes in four seasons hitting within 100 miles of Haiti — how do you then turn around and say that a single strike in one year is a sign of climate change? Obviously, three in four years in one area is *far more extreme*.

So is it ignorance, or outright lies?

I'm not going to name names, but they are revealing themselves with each comment. There is a reason 50,000 Poles turned up in Warsaw to protest the climate conference, because they are calling it like they see it. They know exactly what this is about. That the folks pushing Haiyan as evidence of global warming — they are trying to revert it to climate change, because they have lost on the other matter — would use the misery of a poor people living in an area where this can occur any year says enough. I often think that climate alarmists can sink no lower. But in a year where the *global* tropical activity was — again — well below normal (seventh year in a row) and the Atlantic shut down completely in a situation where it never has before to this extent (warm cycle of the Atlantic), you have to state the facts. The claim that Haiyan is an example of climate change is being made by people who are either ignorant of the facts or simply lying.

And complicit in this is the mainstream media, which are pushing these distortions, obviously ignoring countless examples.

No one is against helping people. I am for helping people in a rational manner — not using deceit or ignorance while capitalizing on the misery of poor people to push an agenda.

A week later I issued a "Haiyan Update."

For the record, Haiyan was the *seventh-*strongest recorded typhoon to hit the Philippines, which has been struck 58 times by super typhoons.

This also shows us that, contrary to what climate alarmists are saying, landfalls have decreased.

By the way, Sunday's tornado outbreak was likely the fifth-largest for November. Autumn is the second tornado season and it correlates most strongly with a warm AMO — the cycle we are in now!

The tropics are a ready-made weapon for these people, and they use it wherever they can. Same with droughts and floods.

I have always enjoyed going after them before the fact, knowing that most of them do not look at the weather until the weather happens. The Texas perma drought started reversing in 2013 and continued through the following years, much like the great drought of 1952-54 reversed by 1957-58. The media have never called out the Texas Climatariat, as Judith Curry might say, that made the forecast for the drought, just like they ignored the perma drought that was predicted for California. On May 22, 2014, I wrote "The Coming Nine Months: Mud in the Eye of the So-Called 'Permanent' Drought?"

My last couple of posts have dealt with major weather/climate stories I think are lurking. The worry about the upcoming hurricane season is something I will have every year for the East Coast until we reverse the pattern we were in during the 1950s. I think we have another five to seven years, and I am very surprised it has not been worse. But it takes just one 1954 or a 1954-55 back-to-back season to get that score close fairly quickly.

I am on this "kick" to expose before the fact some of the nonsense being spewed and accepted by what I consider is an increasingly desperate climate alarmist crowd. It's now or never for them, though, to their credit, they have the political will backing their interests in the form of the EPA, which couldn't care less about the fact the department's lines of evidence have nothing to do with reality.

My one-month electricity bill in January was higher than the totality of the seven months ending in November. And it's going to get worse, as I discussed here.

There were places in the Northeast this past winter that were within a few days of running out of fuel for electricity. At the very least though, I give the president credit for keeping his word on what he was going to do. I believed him when he said, "Under my plan ... electricity rates would necessarily skyrocket," and I certainly believe him now.

I was taught to respect the president, whether I agree with him or not, so his own words and the results I see are what I judge him on in this matter. He is keeping his word with the policies enacted, which are not nearly as draconian as some of his backers want. They are attacking him for not being even more aggressive. Believe me, if you saw some of the things I see, you would actually discover that, as bad as it's getting, it could be — and will be — a lot worse compared to what is being advocated by his supporters. Do you realize North Korea is being lauded as a champion in the fight against "climate change"? Look at this headline out of the Guardian: "North Korea: an unlikely champion in the fight against climate change."

Ever seen a picture of North Korea at night?

The driving force behind all this — the facts on what was said about the actual climate and what is happening — is what I am taking apart piece by piece.

The Arctic ice cap is not gone. Wildfires were again below normal in 2014. Tornado activity was not at a record low like last year, but it's still less than 25 percent of normal. AGAIN.

And now comes the latest: mud in the eye of those trying to say the U.S. is in some permanent dust bowl again.

First of all, major U.S. dry periods are a product of a cooling tropical Pacific. In the decades such as the 1950s through the 1970s, when the tropical Pacific is cooler overall, the U.S. is drier than normal in much of the nation. It is exactly opposite in the years the Pacific warms, which, by the way, correlates nicely to an increase in global temperatures until the atmosphere adjusts to the warming tropical ocean and temperatures level off. But the idea that *global warming* causes drought here in the U.S. *is opposite of the facts!* It's when the Pacific starts to cool and global temperatures start to drop that we see it dry out.

When the tropical Pacific is predominately cold like it was in the 1950s through 1970s, as shown by the multivariate ENSO index, the U.S. is very dry during the growing seasons. When warm, like it was from 1981 to 2005, it's wet.

You can see more dryness starting to show up again as of late, and the drying is starting anew — but so is the drop in the global temperature! It's not global warming causing the reaction that leads to the drying over the U.S. but the lessening of available moisture because the source region for so much U.S. moisture — the tropical Pacific — is cooling. The temperature chart for the last 10 years shows the slow, jagged cooling that has begun after the adjustment to the warming that occurred when the Pacific went into its warm cycle in 1978.

It's clear what is going on. What happened before *naturally* is happening again, as is to be expected given the cyclical nature of the climate due to the design of the planet.

One can see the clear reaction to the warming Pacific, the adaptation (leveling off of temperatures) and then the slow decrease starting as the tropical Pacific turns cooler overall.

During the cold cycles, however, there are warm spikes, and these are the short-lived El Niños that develop in the overall colder periods. We think there is one on the way. There is a lot of talk about the so-called Super Niño coming, but we do not find the physical drivers present for that. **[It did not occur that year. It waited until the following year.]** Quite the contrary: The coming El Niño looks similar to 2002 and 2009 **[which it was]**, and that means that the winter is likely to be cold and stormy, and we are already warning our clients about that. However, the rain issue is huge. I believe that the dust bowl will get mud in its eye the next nine months as much of the drought-stricken Plains from the Texas Panhandle to Nebraska gets above normal rainfall June through March. We believe another great growing season is on the way for the nation's breadbasket, and I believe we will get some relief for California in the fall and winter! Additionally, we are seeing the modeling getting wetter and wetter. Models are tools, and we set our forecast up and then see what modeling agrees with us to fine-tune it. But I like the idea the modeling is seeing, which is what we have been seeing!

Some may wonder, why do I agree with the models now when so many times I am pointing out they are wrong? Because I believe they are right now, since this is what I have been thinking is going to happen based on our research. Therefore, if the models come to us (most of them are), it is a good extra opinion that is backing us up. Models are tools for the answer, and when they can be used, you use them!

I think because we will go back to an overall colder signal next year, this is a one-year event. But the fact is, it will show that it's nature in control, and we are forecasting it beforehand! The climate model sees it too. I might add that in the middle of the 1950s, the drought across the U.S. was worse overall according to the Palmer Drought Severity Index.

It is true California is worse this year. The patterns are similar, but not exactly the same. However, no one in their right mind can say increased CO_2 is the cause of more rain in one place and less in the other. Such events are products of natural variations in what is a similar cycle.

By the way, a lot of rain is being forecasted in California, which will give them a break. But again, this is not over yet; as you saw, this is a multi-decade event that was well-forecasted by a lot of us, though no one would listen 10 years ago. I coined the phrase "time of climatic hardship" and was on "The O'Reilly Factor" with it around 2008 because of what we saw coming with the overall cycle — the similarity to the '50s drought and the worry about hurricanes, which, so far, have not been as bad as I feared. Then again, five of the eight major storms that ran the East Coast between 1954 and 1960 occurred in just two years, so we are not out of the woods by any means.

Now here is something interesting. In 1972, Albert Hammond observed how dry it was in southern California and the occasional major rains that occur and made a hit record out of it with these lyrics:

Seems it never rains in Southern California
Seems I've often heard that kind of talk before
It never rains in California, but girl, don't they warn ya
It pours, man, it pours

It was actually one of my favorite songs my junior year in high school. Perhaps Gov. Jerry Brown, who is really ramping up the global warming rhetoric in spite of some questionable policies that are cutting irrigation to protect a smelt, listened to different music. Wasn't the Jefferson Airplane big around that time? (Ha ha. Even they sang, "When the truth is found ... to be lies...")

When you have rock and roll artists observing nature, then you know it's something obvious.

But here is the problem. We have many more people living in these areas and, unfortunately, this gullible population is ripe to be exploited by those with an agenda not based on facts. It's easy to refute them; you are seeing me do it in the time it takes to write this. Every time something comes up, it takes me only the time it takes to go to the maps to counter it. It's all there. So, as I've previously opined, it's the climate (rather, global warming — they said it, they own it) *agenda*, not the climate, that is the biggest threat to our freedoms. Then again, I believe it was Joseph de Maistre who said, "Every country has the government it deserves."

As in so many things in life, we reap what we sow, and if you wish to simply follow along without verifying, you will go where you will no longer be able to decide for yourself. I never ask anyone to believe me. Go look for yourself ... while you still can.

This piece is called "Disaster Du Jour, Part 1," published on May 22, 2015.

Sometimes I think climate alarmists are like a swarm of locusts, moving from field to field. It's actually a great strategy because there's always going to be some location on the earth where some kind of extreme is occurring — though, in most places, the weather is tranquil the majority of the time. If it wasn't, there wouldn't be so many people on the planet, which I suspect is another agenda that drives all this. Simply find a place where a rare event is occurring, then blast headlines that claim it's because man is destroying the planet.

It's disaster du jour. The West Coast drought is bad, and in some places as bad as we have seen in the time of accurate records. But low lake levels are not being helped by atrocious planning or policies that value smelt over people (example: allowing a river that could be used for farming and civilian water supplies to not be properly utilized to protect a creature called the Delta smelt).

Let's go back three years to Texas and the southern Great Plains and the talk of a new dust bowl. Last year we said the drought was going to begin reversing.

It has reversed and continues to reverse. But do you hear anyone who has turned the focus to the California drought pointing out that the hysteria in the Southern Plains is reversing? The same type of thing happened in the 1950s — hot and dry in the Southern Plains, then relief with the El Niño late in the decade. [**Two years later, the California drought is gone, and the resultant rains lead to extra foliage that fueled a big wildfire season — something we forecasted in the spring!**]

To be sure, the Pacific, which is going through a similar cycle as the 1950s when it suddenly cooled and then warmed for a couple of years, will cool again. And with so many more people living in the South and in California, this is an adaptation-to-nature problem, not a let's-all-act-to-change-nature problem as it's being advertised. But the media blindly follow the radical missive of the climate agenda, so the disaster du jour rules the headlines.

A lot of this is like the ozone scare we hear about all the time. Back in 1992, Al Gore was very concerned about the ozone hole opening up. What puzzled me is that if it was opening again, at some point it had to have closed up. So whenever we hear about the expanding ozone hole, how come no one in the media asks, "Well, why did it shut to begin with? Isn't that natural?"

Same goes for drought, floods, etc.

Finally, check out this quote:

"There's little doubt about our changing climate. The fierce winters of yesterday are disappearing, tornadoes and hurricanes are becoming more vicious and weather trends aren't 'trends' any more. They can't be depended upon. Just about anything can happen — and does."

Sound familiar? Sound like what we are hearing right now? Well, it was from a similar cyclical pattern in the 1950s that lead to the aforementioned fierce drought in the Southern Plains. It's from the September 1953 edition of *Popular Mechanics Magazine*, only then they were blaming it on the Atom Bomb. Now they blame it on CO_2. We know the former was wrong, and it's very highly probable the latter is too. The common denominator in both is the cyclical nature of the climate, inherent in the design of the system.

You can see disasters du jour are nothing new when speculating on nature. What may be new though is the change in attitude. The president in the 1950s was Dwight Eisenhower, who had reservations about the military-industrial complex when he left office. But something that's rarely brought up is this snippet from his speech on the scientific-technological elite:

"The prospect of domination of the nation's scholars by Federal employment, project allocation, and the power of money is ever present and is gravely to be regarded. Yet in holding scientific discovery in respect, as we should, we must also be alert to the equal and opposite danger that public policy could itself become the captive of a scientific-technological elite."

How true it is that the weather repeats itself in the large sense. The weather example from the 1950s is showing us this today. But Ike's words foretold of what we are seeing today in the climate fight from the scientific elite, and they are certainly more accurate than the hysteria of today's disaster du jour crowd.

Interesting. As highly regarded Ike was for his service to his country from soldiering to civilian office, it turns out he was a heck of a forecaster. His words foretold of the disaster du jour mentality that has taken hold today.

Here's another hurricane article that was written in October 2015. It's an annual event — the debunking of hurricane hysteria with history. "Hurricanes Become a Ping-Pong Ball in the Climate Debate":

This caught my eye, from USA Today:

"Study: Climate change adding billions to U.S. hurricane costs."

Oh really?

The study quoted ended in 2005 and didn't factor in the last 10 years. During that time, by way of the Saffir-Simpson scale, there were no major hurricane hits on the U.S. (On my power and impact scale, there have been three borderline majors.) It's been an amazingly quiet period, meaning the dire ideas that we heard about have been nothing but *wrong*.

Another USA Today headline from 2006:

"New study ties global warming to stronger hurricanes."

There are numerous articles on how global warming (climate change is a redundant focus-group-driven term that is now used since there has been no significant warming for nearly 19 years) is causing everything to be worse. Tropical cyclones, since they are awesome to look at and report on, are front and center as examples of how bad things are. But there is a big problem here: *They aren't as bad.* Data from National Hurricane Center researchers Eric Blake and Chris Landsea plainly shows that the busiest decade (2001-2010) for major hits in the last 30 is equaled or exceeded by six of the previous 15 decades!

I often go after Rhode Island Sen. Sheldon Whitehouse — who has been pushing the idea of RICO-like investigations on scientists who do not believe in human-induced global warming — for his pronouncements on hurricanes being worse now than before. It's astounding given he is from a state that was devastated in 1938, 1944, 1954 and 1960 by major hurricanes. But when we look at the hits of majors — 1871-1880, 1891-1900, 1911-1920, 1931-1940, 1941-1950, and 1951-1960 — all were decades equal to or greater than 2001-2010. Furthermore, the 30-year period from 1931-1960 had 61 hits, or two a year, 27 of which were major (almost one a year). By contrast, the most recent 30 years ending in 2010: 43 hits, 19 majors. *Not even close!*

The idea that costs are going up is *not* from increased frequency and intensity of storms. It can't be since the frequency and intensity of landfalling storms has decreased. So we have one side of the debate claiming that increased hurricanes are a reason to suspect there is, as they put it, climate change, even though the facts show there are less landfalling storms now than there have been in many years before.

Right off the bat, the immense buildup of coastal development means that storms are going to be much costlier. I will leave it to others to play with inflation figures, but an example could be Hazel in 1954. This is the latest Category 4 storm to hit the U.S., and that it hit on the coastal Carolinas in mid-October is extreme in itself. In 1954 dollars, the storm did $354 million in damage. The government's inflation calculator says it would be 10 times that now, but crucial is the fact that in 1954 there was not near the amount of buildup in the areas Hazel hit (it had hurricane force winds all the way to Toronto!). This is not climate change or global warming. It's a product of man believing he is in control of the Garden of Eden, as if this is paradise and nothing bad happens. The thumbing of the nose is not CO_2 in the air but buildings on the beach.

My side of the climate debate counters this with what seems to be an intuitive argument about the lack of storms being a sign that there is no global warming. This is a very dangerous tactic. It's one thing to counter, as I did above, the argument that landfalling storms are stronger and more frequent, simply because *they aren't*. But that is all that means. So what happens if seven majors hit in two years like in 1915 and 1916? And guess what? I am very concerned that we are about to see a major burst of hurricanes between 2016 and 2018. Why? because I have seen this before. If we look at sea surface temperatures for next hurricane season, by next July the El Niño is gone and is reversing to a La Niña! The main development region of the Atlantic is very warm. **[Sure enough, 2017 showed what can happen.]**

Major bursts of landfalling storms occurred in '95 and '96 after the El Niño of '94, '98 and '99 after the El Niño of '97, and '03, '04 and '05 after the El Niño of '02. It's natural, it's happened before and it's about to happen again. So I would not be touting the lack of hurricanes as anything but what it is — a lack of hurricanes. But much more deceitful, in my opinion, is using hurricanes as a sign of global warming. It shows the gall of the people suggesting this. Even with facts staring them in the face, they simply ignore them and spew rhetoric anyway.

Hurricanes are nature's way of taking heat out of the tropics and redistributing it to the temperate regions. Weather and climate are nature's way of seeking a balance it can never attain because of the very design of the system. Nothing more, nothing less. Attributing hurricanes to so-called climate change is provably wrong. Hurricanes are much more than ping-pong balls for someone's agenda. They are an awesome display of nature, sometimes resulting in terrible consequences. But they're all part of the natural up and down that is inherent in the system.

"Lessons From Patricia" was written on Oct. 26, 2015. We have recon that can report on storms at will, and so when Patricia went wild in the Pacific, climate alarmists were out in full force. I countered immediately. By the way, notice the line at the end of this predicting an uptick in Atlantic activity.

Failing to bring about the kind of devastation a 200-m.p.h. hurricane does because (a) it did not have 200 m.p.h. winds when it reached the coast and (b) it hit a sparsely populated area, Hurricane Patricia immediately became an example of how hurricanes have become a ping-pong ball between two different climate agenda. As Patricia grew to scary dimensions, the screams grew from what seems to be a cadre of climate alarmists who claim that Patricia, like Irene or Haiyan, is evidence of so-called climate change. (Again, no one denies that the climate changes, it's just the cause that is up for debate.) Interestingly enough, Irene fell apart coming up the East Coast, something that may not happen in coming years. And in the late 1800s, a similar storm to Haiyan instead of weakening went into Haiphong and killed over 250,000 people.

My point is that *every storm* now is being blamed on climate change, but no one is accountable for their ideas when a storm falls apart.

This list should put some of the "it's worse than ever" cries to rest. Example: A flood on the Yellow River in China killed 900,000 people. What's remarkable is that when you look at the top 16 at the end of the file, in spite of better detection methods and more people living in harm's way, it seems like the spread is fairly even.

Amazingly, the very day I wrote the hurricane ping-pong ball article the media played right into the missive with agenda-driven statements on the storm.

I informed people through Twitter that Patricia would be off her peak when she hit: "Patricia should weaken a bit before landfall as monster storms need perfect conditions and drier air may get entrained. Still a beast though."

The point is the weakening was being seen, and I assume by more people than me. Monster storms like this need perfect conditions to remain powerhouses. I use the analogy of the 9.1-second, 100-yard sprinter. Just a tweaked hamstring and he is nothing out of the ordinary anymore. The stronger the system, the more perfect the conditions have to be. So if you "tweak" the perfect hurricane with less than perfect conditions, it will weaken. Patricia "filled" 30 to 50 millibars (rapid pressure rise) as it approached the coast in the last six hours — the *exact opposite* of what happened when it ramped up. The power and impact scale I have takes into account pressure rises and falls. A storm filling (intensifying) more than two millibars per hour gets a category subtracted (added) to it; more than four millibars per hour, it's *two categories*.

This is not my original idea. I learned this method from two National Hurricane Center forecasters, Gil Clark and Bob Case, in the 1980s. The physical reason likely lies with the fact that in a rapidly weakening system, the storm's ability to bring strongest winds to the surface is impeded, since the very reason it is weakening means there are processes occurring to disrupt it! Naturally, the opposite is true. But my power impact scale said this was no more than a Category 3. It rivaled Lili in the Gulf as it approached Louisiana in 2002, falling from a 4 to a 1 in just 12 hours. In that case it was because of dry air and cool water left from Isadore a week or so earlier, but the same thing happens. With Patricia, there is a chance they will find an area that sustained Cat. 5 winds, but it would be a very tiny area. In the large scale, the rapid weakening means the power and impact scale I have, which is meant to give people an idea of the total power of the storm, says this was a major storm but not as extreme as the Labor Day Hurricane of 1935, Camille in 1969, Janet in 1955 and several other storms in the *Atlantic Basin*!

Conversely, storms like Celia in 1971, which went from a Category 1 to a 4 in 24 hours, or Humberto in 2007 can do the opposite.

But here is my argument as to why a hurricane like Patricia *does not support* the argument that storms are stronger. *We just observed a beast over the water, the strongest ever observed in the Western Hemisphere, right?* Well, a lot of Cat. 3s have hit the coast that did not have reconnaissance, so how do we know if they were *not* that strong over the water? We don't. The only true metric is landfall intensity, which we have known through the years because there have been observations on land all that time. So in reality, though powerful, Patricia was just another strong hurricane that hit the Mexican coast. If this were the 1950s, you may not have even known. Speaking of the 1950s, did you know that the only Atlantic basin recon disaster occurred with Janet in 1955, which made landfall as a 914 mb hurricane? The recon went down in the storm, likely because the extremely low pressure of the storm caused its altimeter to malfunction. We had recons, but they were very infrequent, not constant like we see now. Patricia likely did not beat Janet at landfall (it hit from the Caribbean), as the pressure had already risen to 910 mb two hours before landfall and it was filling rapidly. It did not beat the Labor Day hurricane of 1935 with a *verified* land pressure of 892 mb.

Think about that. Patricia was south of Mexico. The Labor Day hurricane of 1935 went through the Florida Keys!

Events like the 1935 storm, Janet, Celia, etc., are a warning that the rapid weakening of Patricia is not something that occurs with every storm. Large, powerful storms such as Katrina and Rita will weaken as much as a day away because their circulations are such that they pull in drier air from land. Both weakened to Cat. 3s from Cat. 5s. But storms like Camille (1969) and Charley (2004), both of which, by the way, occurred in El Niño years, were at their peaks when they hit. The smaller, more intense storms approaching our coasts, which don't have mountains right on top of the beach, do not act the same way as what we saw with Patricia. Moral of the story: Each one is unique to the circumstance it's in. And every example of "worst ever" can be countered with an example of the opposite.

One forecast is already verifying though.

With the Atlantic coming to life in the coming years, look for that missive to grow even more distorted. That's a forecast you can bet on!

Of course, politicians on the Left love using any system as an example of "climate change." But Hillary Clinton's using a run-of-the-mill hurricane at the height of the season as evidence displayed ignorance of fact or deceit (you be the judge).

I wrote "Hermine a Poor Example to Push Man-Made Global Warming" on Sept. 8, 2016. This is a good example of the political nature of this issue.

I read this article that appeared at Climate Change Dispatch with great interest given what I do for a living:

"Clinton says Hurricane Hermine was caused by climate change as hurricane drought persists."

I have watched and forecasted hurricanes on a professional level for almost 40 years. So what Hillary Clinton is asking people like me to believe is that a storm which took 15 days to develop and then hit Florida as a Category 1 hurricane, and then never re-strengthened off the Mid-Atlantic in spite of record warm water, is a sign of extremes. Actually, I am sure she doesn't care that people like me show the facts, since the idea is to get it out there and rely on the idea that the media will gladly pick it up, publicize it and, once the horse is out of the barn, know there is no way to get it back. In reality, portraying Hermine as some kind of climate change demon is either ignorance as to the history of hurricanes or deceit.

The period 1900-1949 was loaded with many additional hurricanes. In the 1940s, Florida was like Grand Central Station for MAJOR hurricanes. And of course, CO_2 was much lower then.

So the question is, why is Mrs. Clinton warning us about something that occurred much more frequently in the past, yet she's trying to blame it on an agenda-driven issue?

By the way, given our hurricane forecast, there are likely to be "better" examples to use later in the year. But they are not supportive of the agenda she is pushing but simply nature being nature. Using a storm that arguably underachieved relative to what more numerous storms have done in previous years as an example of impending climate doom is deceptive and indicative of a person out of touch on this matter. It's that simple.

Of course, it can be argued such things are small beans compared to other issues that bring up questions, but those are beyond the scope of this commentary. Those who have eyes can see the obvious here.

In 2016, a nasty drought was occurring in the interior Southeast. The year before the complaint was about floods. Notice the pattern here — droughts and hurricanes are the big weapons in the arsenal. And since there are always droughts and hurricanes, we are going to hear about this all the time.

I wrote this on Nov. 23, 2016: "The Southeast Drought: Nothing New Under the Sun."

> I realize it's going to be tough to separate the honest climate brokers out there from those who simply are trying to drive home points based on an agenda they cannot retreat from. The classic example is the current Southeast drought. Of course, it's not a drought for those who were in the path of Hermine, Julia, Matthew or the non-named mid-summer beast that hit Louisiana. (Good job Donald Trump on not needing a named storm to visit and try to help out.) The fact is this: Between 25 and 33 percent of summer rainfall in the South (from roughly I-45 in Texas to the SE coast) comes from systems that have some link to the tropics. Since evaporation rates are high, if there are no tropical systems, it is going to be dry. But the averages are not because of even distribution but instead result from back-and-forth swings that contain a lot of dry years and then some big hitters.
>
> Remember the 2004 hurricane season with the assault on Florida and the South? There was no drought in the Southeast then.

However, by 2007 it's quite dry over a large area of the nation, but the turnaround is dramatic over the Carolinas.

But it goes back and forth. Systems do not have to be named to cause problems. Slow-moving troughs will interact with deep tropical moisture, and that is the key **[Harvey in 2017 for example]** — the years in which you do not have stronger than normal ridges over the Southeast allow extra input of tropical moisture. Thus, they will make up for deficits. In any case, last year the indirect interaction of Joaquin and a strong upper trough helped. It became quite wet again.

Last year there was the usual screaming and yelling about climate change when that storm hit the Carolinas, yet such events are part of the natural ebb and flow of nature. Nature needs conflict to resolve the differences that are inherent in the system. I wish the people who use every Tom Dick and Harry event — for Marx Brothers fans, how about Groucho, Chico and Harpo? — would simply look at how this always happens: the pendulum swings.

One more thing. What happened to the Texas perma drought that was all the rage back in 2012? It's pretty wet there still. We said three years ago it would reverse and it has. And guess what? It's going to get dry again, then wet again, then dry again. You know why? *It's Texas. That's what weather and climate do in Texas* (and in most other places too).

Interested parties should understand what all of us old-timers were taught. Weather and climate are merely the atmosphere's eternal search for a balance it can never have. It's because of the nature of the system, which is built for conflict. It's just a shame that not only do some folks have to focus on forecasting these events, they also have to counter missives that are based on an agenda. The true goal should be getting the forecast right and explaining the why before the what.

Another example had to do with Harvey. This was written on Aug. 30, 2017: "Hurricane Harvey: Nothing That We Didn't Know."

There is much being made of Harvey and climate change. Meteorologically, as far as the intensity of the storm, it's tied for 14th strongest. The storm above it is Hazel. Now, let me ask you: Which is the more extreme as far as deviation from normal with pressure, which is a good metric to objectively evaluate how extreme a tropical cyclone is — a storm that hit in *mid-October in North Carolina*, or one that hit the central Texas coast in late August? Let's also look at Harvey in relation to other hurricanes in Texas. Behind it is the 1915 Galveston hurricane. That is the lesser of the two evils, because the 13th right above Harvey is the 1900 Galveston hurricane that killed 6,000-12,000 people. And right above that one is the Freeport hurricane of 1932. *Notice when these are occurring.* Then there is the 1916 cyclone in Texas — just a year after the 1915 Galveston hurricane — and Carla in 1961. Again, this all occurred over 50 years ago. Then there is the 1886 Indianola hurricane. They are all hitting in the area that Harvey hit.

So the question becomes, if those same storms, almost all stronger, from many years ago hit today, would they be a sign of climate change? Why is Harvey — and not to downplay the storm, but it was one of many and less intense than most — a sign the climate is changing, but these other storms would not be?

Harvey got trapped not by an expansive subtropical ridge, as one notable climate scientist claimed, but by an abnormally large-scale trough over the eastern U.S.

That lined up in textbook fashion with Phase 2 of the MJO that was occurring. I jumped all over this on Aug. 21. This is a hyperactive phase in the tropics near the U.S. and also a cold pattern over the U.S.

The development of Harvey and the pattern that caused it to stall were all on the table well beforehand for those who looked. However, if you did not look, or you were unaware of previous metrics of strength, then you would fall for the arguments that this is part of climate change. All those other storms continued moving, so they went inland and died as they got farther away from the ocean source. Because either a warmer than normal or normal pattern steered them that way. Harvey stalled because of a pattern that has happened before, was on the charts, and involved the opposite of arguments that would lead to a warming conclusion. The stalling of the storm was key since lesser storms that have stalled in Texas have dumped almost as much rain, the most notable being Amelia in 1978, a mere tropical storm that dumped 48 inches of rain. Naturally, a stronger stalled storm would dump more rain, but what stalled the storm was not a result of climate change but rather a well-known, well-forecasted pattern. So if the 1935 Labor Day hurricane — the most powerful storm to be recorded hitting the U.S.; a storm that went from a tropical storm to a Cat. 5 in 36 hours — occurs again, why would it be climate change now but not then?

If the 1938 storm — a storm that took down two billion board feet of trees in New England, caused major river floods in western New England, flooded Providence with 13 feet of water in a storm surge, and had a wind gust of 186 m.p.h. at Blue Hill — occurs again, why would it be climate change now but not then?

If Donna of 1960 showed up again — with hurricane force winds in every state from Florida to Maine, never recorded before or since in U.S. history — why would it be climate change now but not then? **[Irma did show up and was not as strong as Donna, but climate change was blamed anyway.]**

I can go on and on with countless storms.

The answer: It is nature doing what nature does. And coming out after the storm and claiming it's something else reveals either ignorance of the past or, if you do know, an agenda based on deception. If I saw the people commenting on this now making a preseason forecast, or even five days before when the obsession was the eclipse, then perhaps I would be more open to those ideas. But telling people why after the what is Monday morning agenda-based quarterbacking. Perhaps that is the lesson of Harvey.

After Irma I wrote "The Absurdity of It All," which sums up the nonsense of using two storms — or any storm — as evidence of man-made global warming.

> We have had two major hurricanes strike hurricane-prone areas back-to-back in the past two weeks. The record, by the way, is 23 hours in 1933. I am still waiting to see if we had verified Category 4 sustained winds with Irma, but no matter — Irma and Harvey were major hurricanes. These areas have been left alone, along with the rest of the nation, from the known extremes of hurricane frequencies for the past decade. That in itself should raise questions as to the idea that CO_2 is causing storms to be stronger and more frequent. A three-week, well-forecasted period (as far as the hurricane spurt) should immediately have people calling into question the shrill voices we are hearing now.

> Think about this. Suppose the system that ended up re-intensifying into Harvey had just kept moving. Suppose we had a normal subtropical ridge like we did with Carla — a much bigger storm in 1961 for the Texas coast as far as the extent of wind and lower pressure are concerned — instead of a pattern with a *weaker ridge* that allowed the trough to catch and stop Harvey and produced tremendous rain. It would have been a major hit, but missing Houston and Corpus Christi and ending up like a lot of the other storms that went into that area of Texas. This is known to people like me who study these but not to most people. Harvey's infamy is because of circumstance — opposite the man-made global warming missive about stronger ridges.

Now consider this: *What if the ridge over the Southeast was stronger and forced Harvey farther south across the southern Gulf?* There would be no time for it to intensify, no time to come up, so it would have been like Franklin and Katia — rather normal western Gulf of Mexico hurricanes that are expected over most seasons. A stronger ridge may have suppressed Harvey! The weakness there is what helped him gain latitude. But my point is these occurrences are pattern-driven, not CO_2-driven. Harvey was a disaster because of reasons that have nothing to do with any argument about climate change. One could even argue that the stronger subtropical ridge could have saved the U.S. from Harvey, but, alas, there was no stronger ridge; there was a trough. So instead of moving west from a more southerly latitude, it came up.

A few weeks earlier Franklin in the southern Gulf had missed the connection, because there was not the same pattern, and so it went into Mexico. The pattern, by the way, was well-forecasted and a well-known pattern for U.S. landfalling hurricanes!

Now back to Irma. Let me play the opposite card, the more intense card. Irma's path 50 miles south to the north coast of Cuba did two things: 1) it stopped the intensification process, and 2) reversed it. If the storm was 50 miles further north it would have been over the water. We know the atmosphere and sea surface temperatures were ready for the deepening because it was starting to intensify before reaching Cuba. And after it came off Cuba it intensified again for about 12 hours.

But that 12 to 18 hours over land not only took away the period it would have been intensifying, it made Irma re-intensify at a weaker point. The result was a big mitigating circumstance, one I was explaining on national outlets. Fifty miles likely saved $50 billion, and 100 miles $100 billion, because a track away from Cuba and 100 miles to the east would have been a worst-case scenario for southeast Florida.

Donna in 1960 was far worse than Irma for Florida, as it hit all the way up into Naples as a Cat. 4 (Irma may not even have been that in the Keys). *Donna's sustained winds were higher than Irma's gusts!* The 1935 Labor Day hurricane went through the keys with 210 m.p.h. winds. But Irma is worst ever?

Was it CO_2 that "saved" Florida and forced the storm south into Cuba, or was it the *totality of the forcing* that has always been there? How can one tell if CO_2 caused Irma to be a monster in one place but less of a beast in another? The fact is, some modeling caught the move into Cuba, and no model I know of has a CO_2 input. The argument that the water is warmer because of climate change doesn't hold water because we have had stronger storms in there (the 1935 Labor Day storm was 35 mb lower and 50 m.p.h. higher than this), so preexisting knowns debunk the argument of the unknown.

It really is getting absurd. That these missives are being used at a time of tragedy to push an agenda with an "I told you so" attitude based on something that a) just looking at history showed was bound to come back and b) was warned about by our forecast team before this season using analogs of other seasons where CO_2 had nothing to do with the impact makes it more absurd. And the same people who are trying to shut down debate are in reality the ones crying fire in a crowded theater to create chaos.

So here is the rule for agenda-driven comments when a storm is coming: If it hits someone, it's climate change. If it misses or is way out to sea, it's weather.

The truth: It's all weather, and there is nothing new under the sun.

There are other examples. In fact, they are so frequent I can't write on them all. But in the weaponization of weather, there are three factors that enable this tactic to be used.

1) The people pushing this believe they are heroic revolutionaries. They have set themselves up as the underdog, creating a straw man that they must beat. It is because of the bigger agenda. If you want to see proof of this, a simple look at the frightening quotes of people driving all this is required.

2) There is a willing media pushing this on people who do not have the time or the inclination to actually look at the entire picture. It is lazy journalism at best and, like we are finding out in other aspects (Harvey Weinstein, for example), the media will bury anything that does not fit their missive.

3) The dispensing of information and the ability to see everything and make it seem like it's common (such as the reporting of weather events) is very powerful. A tornado in the middle of nowhere can be captured easily on an iPhone. Storm chasers are getting to

places they never would have been able to get to before. More people are living in harm's way.

In the end, the old weather adage that anything can happen and probably will is the basis for the weaponization of the weather. They know extremes will occur, so they simply use them.

What is so disingenuous is the fact that there are some things going on that do raise questions, but they are either unaware of them or are ignorant of the actual implications. I have blogged, for instance, on the antics of the Madden Julian Oscillation. I can counter any storm or event my opponents throw at me, but when a large-scale event like the MJO does something not seen in 45 years, then one has to acknowledge that it's a big event. This is probably the first time you have heard this MJO revelation, and it's likely the first time too for many on the other side reading this. That simply tells me that they don't really know what they are looking at and are simply using weather events to push their agenda rather than truly examining the evidence.

When you are looking for the right answer, you look at what can counter your argument, not simply push out narratives that mean little in the entire scheme of the matter.

What is the solution? There may be none. I said in the introduction that I have no illusions about stopping all this. In a situation where truth does not matter, the end is not reliant on truth. I am simply showing examples for the reader so that he or she can make up their own mind as to what is going on. In the end, you have to take measure and decide, but these examples are but the tip of the iceberg. What's under the water is where the majority of the mass lies, and, like an iceberg, what I am showing you is just a small part of what is actually going on. But that is for you to decide.

Chapter 4

Dragging the Weather Into Politics

If you wish to ruin the purity of the weather and climate, drag politics into it. As I stated in the introduction, you must understand my fight is not to win some war but to defend what I know and love. It's why older meteorologists like John Coleman and Joe D'Aleo (and many more) question so much with no funding (not that we need it since the information is there). But with the political machines involved, it's just another way of dragging what should be a source of awe and challenge into the mud.

My first political writing on this was on Aug. 23, 2012: "Mr. President, Tear Down This Wall." It was addressed to President Obama.

> The wall I refer to is one of either ignorance or deception that is surrounding you on the issue of anthropogenic global warming. Given the facts, we can draw no other conclusion. The people around you are either ignorant of the facts or they know them and are being deceptive. Since this issue is a huge reason for why our nation's freedoms and security are in jeopardy, you are not immune from the blame.
>
> There is no better example of this than your blaming last year's Texas drought on global warming. The drought was forecasted by myself and others as soon as the Pacific changed to its colder cycle in 2007. On

national TV and in blog after blog, references to the 1950s were used to describe what was about to occur. Yet the wall of ignorance or deception resulted in a completely non-factual statement, and in doing so, you degraded the governor of Texas to score cheap political points.

Had you known what you were talking about — had you checked the facts and had you torn down the wall — you would understand that the warm state of the Pacific in the '80s and the '90s led to *above normal precipitation* across Texas and indeed most of the nation!

Had you torn down the wall, you would understand that a warm PDO (Pacific Decadal Oscillation), *a natural occurrence*, results in more moisture and a more active weather pattern for enhanced precipitation in the United States. You would understand that this would lead to a more favorable climate for plant growth in the U.S. If you had torn down this wall, you would know that the opposite cycle, the cold PDO, leads to the opposite effects. You would know the U.S. dries out as it did in the 1950s.

The dryness that is now occurring is because of the physical realities of the weather pattern, not what your advisors have been telling you.

Even your own government scientists have come around to the idea we warned about several years ago.

"We can now be more confident that the models are correct," says National Center for

Atmospheric Research's Aiguo Dai, "but unfortunately, their predictions are dire."

In the United States, the main culprit currently is a cold cycle in the surface temperature of the eastern Pacific Ocean. It decreases precipitation, especially over the western part of the country. "We had a similar situation in the Dust Bowl era of the 1930s," said Dai.

He is simply forecasting what happened in the 1950s also (the cold cycle of the PDO) because of the same pattern. This was foretold by many private sector climate realists several years ago and is opposite of what you believe.

You would also know that *global* temperatures would stop rising and begin to cool. So, Mr. President, tear down this wall and look at the temperature trend and how it leveled off compared to the 1990s.

Mr. President, tear down this wall and look at the actual global temperatures.

Mr. President, tear down this wall around you. Your policies are based on information that is being proven false each day. *But you have to look*. You have to open your mind to the facts. This is not serving the common good of the nation you lead.

The policies you are pushing are destroying any chance this nation has of becoming self-sufficient. By handcuffing the nation's economic engine, we are entangled in a morass and weakened to the point that we are in danger at home and abroad. You speak of high-minded ideas. You want to bring back the manufacturing sector. How? With expensive energy and prohibitive regulations and taxes?

As patriots, we beg you to tear down this wall that has trapped you and trapped our nation. A booming economy helps all of us. Imagine unemployment at 3 percent and $1.50 gas at the pump. We cannot do this by handcuffing the lifeblood of the nation's economy with energy policies based on falsehoods about the environment.

The fate of the nation hangs in the balance.

Mr. President, TEAR DOWN THIS WALL.

I got a kick out of the idea that President Obama had an Abe Lincoln-like team of rivals advising him. The only rivalry was over how to get draconian decrees most quickly enacted. President Obama, as he got deeper into his second term, realized no one would step up to him, so, as he famously said, he had a pen and a phone, and he used them. In the run-up to the election I wrote the "Man-Made Global Warming Voters' Guide," published on Oct. 12, 2012.

As you prepare to vote this election season, if it's the disasters from CO_2-induced global warming — I mean climate change... oops, climate disruption (all of the above) — that are of most concern, there are four main points to consider:

1) There is a disconnect between CO_2 and global temperatures. There is no linkage in the entire geological history of the earth.

2) Hurricane hits in relation to CO_2 in ppm are down.

3) The number of strong to violent tornadoes has decreased.

4) The U.S. temperature from 1960-2011 versus the last 30-year average shows a leveling off of temperatures.

So, if the anthropogenic global warming issue is your number one concern, these are just some facts to help you make up your mind. If disaster is imminent based on all of this, then I have some beachfront property at the base of the Rockies I would like to sell you.

In both of the blogs I simply wanted to draw attention to easy-to-understand counterpoints to the political missive. It's interesting that people refer to this as cherry-picking, yet in science, if one thing can prove a "fact" wrong, even if it's just one example, then it is not a fact. For instance: A 15-pound bowling ball, if dropped, will always fall, so you better make sure your foot is not in the way. That is settled science. It happens every time. The fact is that the counters are an orchard, not a single cherry. But tit-for-tat battles do not result in a win, and in the so-called climate war (again, a laughable term), a standoff while our nation's way of life is downgraded — the real mission here — is fine with them.

Fresh off his re-election victory, President Obama went on a predictable climate offensive.

So on Nov. 15, 2012, I wrote "A Response to the President's Comments on Global Warming."

During President Obama's recent press conference (his first in eight months), he said, "We can't attribute any particular weather event to climate change. What we do know is the temperature around the globe is increasing faster than was predicted even 10 years ago."

I must admit, it is refreshing to see the president discard some of the hysteria surrounding Hurricane Sandy and its supposed link to global warming. Anthropogenic global warming activists will attribute every extreme weather event to global warming, which has now been termed "climate change" since the earth has stopped warming.

However, the rest of his comments are certainly a byproduct of what I opined about in my August article, "Mr. President, Tear Down This Wall." He is surrounded by a wall that does not allow the truth about climate change, or at least the light of debate, to ever come forth.

The fact is, temperatures have leveled off, and not only is the temperature no longer rising, but one can also see the disconnect with CO_2.

As far as temperatures rising faster than forecasted, a couple of points:

- The only way that could happen is if they were being forecasted to fall, since they are not rising.

- When you compare the observed temperatures of the past 10 years against all the climate model predictions, the result should do more than raise eyebrows about how much taxpayer money is being wasted on climate science that is proving to be wrong.

Before we adopt carbon policy that can hurt our already struggling economy, there needs to be an unbiased debate about what is actually driving our climate. A leader truly interested in the good of his nation would allow the lifeblood of the economy, which is still fossil fuel energy sources, to flow freely while cultivating the alternative energy sources in a way that someday they can compete and drive down prices further. If the president is serious about doing what is in the best interest of the nation, then he will acknowledge the reality of the global warming situation: It has paused, and in the coming years it will be proven to be cyclical in nature as cooling becomes more established.

I have never made a secret of my New England "JFK" Democratic upbringing. In fact, when I was five years old, I was the entertainment at my parents' parties. I would get up on a hassock at 7 p.m. — my bedtime on nights when people were over — and recite JFK's immortal line: "Ask not what your country can do for you, but ask what you can do for your country" (it seems so long ago and it's certainly not a safe space, victim-driven statement).

Right after President Obama's second inauguration, on Jan 24, 2013, I wrote, "...but Ask What You can do for Your Country."

The president's Inaugural Speech contained many references to climate change and how he plans to combat it.

Is that what he really wants to do for his country?

Has he, or you, seen this UK Daily Mail headline?

- "Forget global warming — it's Cycle 25 we need to worry about (and if NASA scientists are right the Thames will be freezing over again)."

I didn't think so. And I am sure the people driving the global warming alarmism train haven't either. Or if they have, they don't want you to.

Or how about this headline?

- "Latest Empirical Evidence: The Abysmal Prediction Failure Of Taxpayer Funded CO2-Based Climate Models" —C3 Headlines

No, these did not make the mainstream headlines. Nor did the record Jan. 1 U.S. snow cover. Nor the record draw on natural gas for the week of the New Year because of the cold that had enveloped much of the nation. But if anything is a record (or portrayed as a record) the other way, you can bet it's mainstream news.

Google "global cooling" sometime and see how much is out there about this potential problem. And yet we are served up a menu of global warming talk and what to do about a problem that is not there.

Mr. President, did anyone show you any charts comparing global temperatures and CO_2? They show no linkage.

Or the horrid performance of the models versus reality?

These are facts that should make anyone say, "Wait a minute, this is very different from what I have been told."

The sun, of course, is the ultimate driver of the whole system. That we simply ignore its ups and downs is arrogance or foolishness. There is no in-between answer on its long-term effects on the planet. Pretending that curbing the amount of a gas that comprises 1/400th of all greenhouse gasses (water vapor is the most prominent) and 395 ppm of the entire atmosphere (that would be like paying $395 on a million-dollar income) challenges common sense. The idea that CO_2 is absorbing enough radiation to then re-radiate it, warm the air and oceans, and change the climate is in even more trouble, for a sleeping sun means there is less incoming radiation. But think about how incredible is the idea that the absorption of radiation by *1/400th of the greenhouse gasses* would then control the entire vast climate system. It is a reach well beyond reason. And the facts prove it — CO_2 is going up and the temperatures are not.

But I don't think the sun is yet a factor, though I am becoming very concerned it soon will be, and the talk of us returning to the temperatures of the Victorian era are not that far-fetched. For my part, simple oceanic climate cycle theory explains perfectly what is going on. The Pacific started its warm cycle at the start of the satellite era in the late 1970s. The earth was colder because the Atlantic and Pacific had been cold. The Atlantic started its warm cycle in the mid-90s. The Pacific flipped into its cool cycle again in 2007. The Atlantic will do so around 2020. The warming that was observed is perfectly consistent with the addition of heat by the warming of the Pacific and the Atlantic. Once the increase in heat from the oceans (which have 1,000 times the heat capacity of the atmosphere) was absorbed by the atmosphere, the temperatures leveled off. The recent cooling is because the Pacific has flipped into its cool cycle, and once the Atlantic does, this should return air temperatures to where they were in the late 1970s. This is not rocket science—it is a longstanding theory that is ignored today, and with good reason. If correct, it is a huge threat politically, economically and academically to positions that evolved as the warming started taking place!

Quite frankly, the CO_2 theory of global warming is akin to turning on a hot shower in your bathroom, then claiming what warmed the bathroom was you turning on the light. There is a test in front of us, since we are turning off the hot shower (oceanic warm cycles), and yet the light (CO_2) is still on. So, the forecast from me, which was first issued publicly in 2007, is that we will return to the temperatures of the late 1970s by 2030. The sun is another, and perhaps bigger, problem. A colder planet would lead to untold misery given the current policies geared toward the opposite.

What would I say to the president? "You have been compared to Lincoln and Kennedy, among others. Lincoln had a team of rivals. The only rivalry in your administration on climate change is how to implement draconian ideas, not whether they truly have merit. As for John Kennedy, as a confessed ex-Democrat and John Kennedy fan, I am taking to heart the second part of his famous challenge: 'Ask not what your country can do for you, but ask what you can do for your country.'"

What can I do for my country? Fight against a fool's errand that has no value in advancing the high ideals and standards this nation represents. That is what the climate fight is really about. The question for the president in this matter is this: Will you allow true rivals to open your eyes to the mountains of evidence against climate change? It's only then that he will be able to answer the question: What can you do for YOUR country?

Of course, it was futile. When politics is involved the only truth that matters is the truth that matters to you. The other side need not apply.

We had a brutal 2013-2014 winter and I forecasted it. The usual suspects showed up with the idea that the warming of the globe meant it would be colder except when it was warmer. The president blamed the lousy economy in the first quarter of 2014 on wintry weather. Too bad he didn't hire me as his consultant. He would have had the country ready in the fall. On Feb. 21, 2014, I wrote "A Proposal for the President."

I was watching the president in California announce a billion-dollar fund to combat the effects of climate change, and guess what? Every hair on my neck did not stand on end! In fact, I agree with this. Why? Because the climate is changing — always has and always will. We are putting more people in harm's way every day. So we should be putting money toward adapting to the effects *of what comes naturally!*

I grew up on the New Jersey shore understanding that it was just a matter of time before the glorious beach was going to be severely rearranged by nature. Why? It's happened before and will happen again. In fact, in a talk I gave to a reinsurance group, I showed the "Sandy Scenario" and said it had to happen, and I expected it soon (I wonder if that is why the group invited me back to talk again this year). A focal point of the hurricane talks I have done for many years now is something I call the "Philadelphia Story" — a landfalling hurricane from the southeast with landfall on the Delmarva Peninsula, shoving the storm surge up Delaware Bay, while flooding comes down the Delaware River from heavy rains. The intersection of the two is over the ports of Philadelphia and Wilmington. My argument is not why should it happen *but why shouldn't it?* This would be far more destructive than Sandy, as it would mean severe damage in a huge port as well as a much greater storm surge into Atlantic City, which was spared the worst from Sandy as the eye hit the city, allowing the strongest surge to pass by the north. The point is that all of this has been seen for years, and not just by me, but by many, and it had nothing to do with CO_2 levels. Adaption measures are fine. They did it at Galveston after big hurricanes wiped out that city. They did it in New England after hurricanes in 1938 and 1954 devastated the cities of Providence and New Bedford. I have no problem with *adapting* to what comes naturally.

But here is the problem: All the other things going on to back a false cause — to try to prevent what comes naturally — is costing this nation time, treasure, blood, sweat and tears — the latter three because of people fighting to make a better living for themselves, while the noose around the neck of the economic lifeline of the country closes. This is brought about directly by government policy due to the notion that man is somehow destroying the climate.

And here is where the president and I part company. I believe if we quit throwing money at a problem that doesn't exist — CO2 — and put it where it does exist and always has — man adapting to nature — then we have common ground.

An ounce of prevention (in this case, adaptation) is worth a pound of cure. The problem is that their agenda is trying to cure something that isn't there and throwing many more times the amount of money at it than what adapting would cost.

The latest example being touted as a CO_2-induced man-made problem is the California drought. **[It reversed in 2017.]** Never mind we have a state that has Lord knows how much revenue in oil sitting offshore and deflects rivers that could be used for farming and irrigation along with drinking water just to save a smelt. And never mind there have been numerous droughts over longer periods of time when the population was far less. Here is the inconvenient fact: When the Pacific Decadol Oscillation (PDO) is warm, the globe warms and much of the U.S. is wet, California included. Why? More warm water in the tropical Pacific means more moisture available for the U.S. When the PDO turns cold, global temperatures begin to drop. Guess what happens to the input of moisture into the U.S., beginning with California? It's not rocket science.

I am not a genius by any means, but how is this so hard to see? And why in the face of such an obvious link would you not expect what is happening to happen? Especially when you consider that it happened before with the last cold PDO cycle.

[The Super Niño in 2015 and 2016 set up a pattern of enhanced moisture. Therefore, when we had the strong trough develop off the West Coast in the winter of 2016-17, the drought reversed. Another climate alarmist agenda busted, similar to the Texas perma drought bust.]

Many of the "scientists" who are pushing the man-made global warming issue fancy themselves also as philosophers and ethically enlightened people. I am not going to argue that point, for who am I to judge? But I want to ask them this question: What do you think someone like Aristotle would think about this whole argument? I think all would agree that Aristotle is pretty much a heavyweight as far as being a well-rounded scientific and philosophical light. Using his argument about the search for the unmoved mover — the source of what starts a given set of events (of course Aristotle, and later Aquinas, were both talking in realms far greater rather than a silly climate fight) — how much sense would it make for CO_2 to be the unmoved mover, which, comprising only .04 percent of the atmosphere, only has $1/1000^{th}$ the heat capacity of the oceans? Isn't that a bit backwards? I do believe in David and Goliath, but CO_2 is no David, and the entire system is far beyond Goliath in comparison.

Now let's get this argument down to the bare bones. I believe the unmoved mover of weather and climate is the *design* of the system. Of course, after that we can start arguing over how the design got that way — random chance or planned order? But given the length you are going to push CO_2 as something we must control in order to command the climate, your argument would have to be that CO_2 is the unmoved mover. You can't have it any other way, for what you are telling us is that the very fate of the planet is in the hands of this gas, even though there have been ice ages at 7,000 ppm (it's currently at 400 ppm).

Though I lack the advanced schooling of many antagonists on this matter, if the referee was Aristotle, I am very sure I would carry the day.

Logic and reason oftentimes don't exist in politics. Interestingly enough, a lot of left-leaning politicos today are of the generation that came out of the '60s and '70s. I used to love Jefferson Airplane. Remember the opening line of its song "Somebody to Love"?

When the truth is found/To be lies

And this line from "White Rabbit"?

When logic and proportion have fallen sloppy dead

Well, it applies to a lot of the agenda-driven ideas of climate alarmists.

A couple of weeks later, on March 5, 2014, I wrote "Climate Agenda, Not Global Warming, Our Biggest Threat."

I want you to notice how, despite my disagreements, I showed respect for President Obama in all my writings — a far cry from what I see today (like whenever someone the Left doesn't like is in office).

> Anyone who has risen to a position of authority via the electoral process has my respect, whether I vehemently disagree with that person or not. The fact that they must somehow garner the votes needed to be elected is something I could not do. In spite of my desire when I was young to be in politics, I realized very early on, to quote the immortal Ralph Kramden in "The Honeymooners," "I gotta big mouth."

> Again, one must smile at all of this at times.

> But let me put this idea forward. If one goes back and looks at NOAA's longer-term outlooks for last spring, summer and this winter, one sees the lack of the kind of cold that has occurred. Last summer the worry about yet another scorching hot summer — which Weatherbell.com did not have in the nation's midsection and does not have again in the heartland for this summer, where a lot of crops are grown — was in the forecast from many sources. The February-April period last year was uncommonly cold. The winter had "equal chances" in much of the nation. My point is this: The government agency — despite a full year now of colder than normal temperatures and a global temperature that is

going nowhere and is actually drifting down as we have shown countless times — had little visible idea that this was going to happen. My statement back in January saying this winter would have the economic impact of a major hurricane on the U.S. coast is basically what you are now hearing. The winter has caused the economy to be less vital if you support current policies, or if you don't, it's enhancing the disaster that this is being portrayed as.

Strange bedfellows here though, as the people in the top echelon of the government whom I debate global warming with can actually say that what I am pointing out is a reason for why the economy is not better. My counter is, well, why did you not at least ask for some private sector opinion as to where this is going? Certainly, other opinions in both the short and long term could serve you well. There are some of us who in spite of disagreements can try to work with people who don't see eye to eye with us without "selling out" or being soft.

Now let's look at another issue. A few weeks ago, a Ukrainian wrestler was training in my neck of the woods and I got to talk to him about the situation in his nation. He told me, "We want what you have: freedom. That is all." Very simple. Now there is no way I am advocating armed intervention in the Ukraine. But consider this: An America weakened by the shackles of the climate alarmist agenda is no help at all to people yearning to be free. If not us, then who will stand for these people? Russia needs the price of energy to be higher to sustain its goals. By not expanding our ability to drive the price of energy down, because we are burdened by an agenda that claims the number one threat to mankind is global warming (I would suggest it's still the same as it always has been: tyranny over your fellow man), we are powerless to drive the price down and affect this situation. So, you make the call. Which is the biggest threat here to the freedom of man and his advancement or, dare I say, his *progressive* movement toward a better world — global warming, or a global warming agenda that limits what freedom-loving people can do and invites tyranny to crush the seeds of Liberty wherever they dare sprout?

I have not heard from my wrestling friend since he went back to Ukraine. One thing I learned as a wrestler at Penn. State was that one must be strong on your own to be strong for others (teammates) and that the rise of the individual would lead to the rise of those around him who wished to become better. It seems to me that a stronger, more energy-independent America would be in a better position to help people in need. It really is too bad. The climate alarmist position — which says this is such an urgent problem that we must immediately spend hundreds of billions of dollars on alternative energy methods that can barely cause a dent to the needs of the world like current energy sources and nuclear power can — is illogical. Furthermore, we sacrifice trillions of dollars that could be invested elsewhere (like promoting Liberty or even, let's say, developing alternative energy in the coming years) by importing foreign fossil fuels instead of producing it ourselves and paying half the price for it. And looking at the results of the climate alarmist agenda and the misery that it is causing, be it directly or indirectly, it makes no sense at all.

Actually, it does make sense when you add it all up. When you have people who believe a strong America is detrimental to the world and that you are a global citizen first and foremost, it all makes perfect sense. It's political. So not only are policies enacted that cause suffering, but the weather and climate, used as a means to an end, suffer too. Again, if you are in love with something, do you not defend it?

Al Gore has been out of politics for years, but as a former politician he is visible on the scene. In a way, he is simply amplifying his victim status (the man who got screwed out of being president) to a higher calling where the entire human race is a victim of people who want to get rich. Never mind he has gotten unbelievably rich in becoming this heroic straw man (it's a side perk — you get to be a hero and get paid). I wrote "Two Simple Question for Al Gore" on July 16, 2014.

> Al Gore is at it again. He was just in Australia (he should hire Weatherbell.com to help him avoid the "Gore Effect" — Brisbane recorded its coldest temperature in 103 years during his stay) and told BBC, "This [climate change] is the biggest crisis our civilization faces."
>
> A statement like that, which echoes much of what State Secretary John Kerry says, is very serious indeed, so perhaps Al Gore should answer a couple of basic questions.
>
> Some points first. CO_2 in the atmosphere is portrayed in proportions that distort it in the same way a picture of an ant under a microscope would distort its size in relation to the environment around it.
>
> Next, CO_2 has no linkage to the globe temperature in the geological time scale or in

recent times (as the Pacific started to cool, so did temperatures).

The CO_2 graphics are on a scale similar to showing an ant under a microscope and then claiming these monster creatures are taking over the earth. The correct scale of CO_2, since it is measured in parts per million, is to show it on a scale from zero to a million instead of the way it is portrayed most commonly, which makes it look like increases can take over the world. It is 400 parts per *million* total and increases 1.8 parts *per million a year.* It is impossible to create a chart like that in millions, and CO_2 would not show up anyway because its contribution is so small.

The current percentage of atmospheric CO_2 is equivalent to picking 43 people out of 107,000. The yearly increase from the U.S.: 1/4 to 1/3 of a person in that crowd.

Consider this: The yearly increase in the level of CO_2 from all sources is 1.8 ppm. There are arguments as to how much of this is due to man. To make sure that I give my opponents the benefit of a doubt, I will assume *all of that increase is because of man.*

Now remember, the heat capacity of the atmosphere is only 1/1000th of the ocean's. That means when we are talking man's input of CO_2 into the entire planetary climate system, the fact is the part we put into the air only has 1/100th of the greenhouse gas effect, the primary one being water vapor. Water vapor makes up just 4 percent of the atmosphere, and the atmosphere only has 1/1,000th the heat capacity of the ocean. Common sense reasoning shows that the effect of CO_2 has to be boxed in by all this. But let's continue, shall we?

The EPA estimates that the U.S. contributes about 1/5 of the CO_2 man emits, which would be .20 x 1.8 ppm, or .36 (that's *point* 36) ppm. I am not going to use smaller estimates of the U.S. contribution, which are as low as 10 percent. As I said, I am assuming *all the increase* is from man, which is also arguable. But I want to consider the worst-case scenario.

So let's keep this short and sweet. Two questions for Mr. Gore:

1) What is the perfect temperature for the planet?

2) Do you really believe that the U.S. contribution of .36 *parts per million* of CO_2 has any provably measurable effect on weather/climate?

By the way, it's early 2018. He still has not answered either question.

On June 1, 2016, I wrote "Pure Political Theatre, Based on an Agenda." In it I showed our hurricane forecast and where we thought the bulk of the intensity would be, and of course later in the season both Hermine and Matthew showed up. Hermine was already used as an example of the weaponization of the weather, but it also was an example of the politicization of the weather in that Hillary Clinton used a run-of-the-mill hurricane that was more remarkable for how it took over 10 days to become a storm at the height of the season as a sign of how bad things would get. If we could get off with only Cat. 1 hurricane hit at the height of the season, then things would not be bad at all. Anyway, to the blog.

My seasonal hurricane forecast first went out in late March. And with the 2016 hurricane season underway, it's worth taking a look at what we settled on in mid-May.

We are calling for a slightly above normal number of storms — including majors — and ACE, which is an index that attempts to quantify the amount of energy in a given season. The normal is around 104. We are forecasting between 105 and 135. The big worry this year is not the total number but rather the threat of storms developing and intensifying rapidly close to land. I came about all this because of longstanding forecast techniques I have used over the years. Our forecast is not meant to scare people or hype tropical concerns. It's an attempt at helping our clients out with the obstacles they may be facing from the upcoming hurricane season and then showing the public what we are doing.

I have written numerous times on hurricanes and how an ill-informed public are being fed a bill of goods by people trying to claim that CO_2-driven hurricanes are going to be worse than ever. Of course, just like the Macy's Parade is the predictable open of the holiday season, the warnings about the dire consequences that face us are coming fast and furious on June 1, the opening of the hurricane season. This week President Obama had this to say:

"All of us have seen the heartbreak, the damage and, in some cases, the loss of life that hurricanes can cause. And as climate continues to change, hurricanes are only going to become more powerful and more devastating."

The climate continues to change. What, *back to where it was before*?

There is common ground between the president and myself with his first line. But the political theater starts when he makes a baffling comment by inserting the climate change issue: "more powerful and more devastating." *Like what has already occurred before, when CO_2 was much lower?* For the record, 100 percent of the facts show that major hurricanes hitting the U.S. from the 1930s to 1950s were worse than any other 30-year period. That is to say: If the hurricanes are as bad as what I am worried about this year or in the coming years, it will just be a return to where we were in the 1930s, '40s and '50s.

Between 1954 and 1960, there were seven major impact storms in the Carolinas, five of them in 1954 and 1955 alone! And New England — home of Sen. Sheldon Whitehouse of Rhode Island, who apparently doesn't know how bad his state was devastated in '38, '44, '54 and '60 by major hurricanes since he says it's worse than ever now — was hit by three majors in seven years.

Since I am a man that has an open mind, I will assume the people pushing this do not know this. I am assuming the president can understand there is common ground to be found among people who love their beaches and have put so much in harm's way and worry that a return to what hurricanes did in previous decades would be devastating. But that has nothing to do with climate change unless you are saying the climate will change back to where it was when these hurricanes were pounding the U.S. It also has nothing to do with CO_2, since it was lower. So the climate change rhetoric is political theater and is uncalled for.

After all, why would they say it's going to be worse than ever if they are just saying it's going to return to levels that were not seen in 50 years? If it happened before, why is it now because of CO_2? The answer: *It's not.* It's because of long known physical drivers that set the pattern up that are conducive for an enhanced threat. I for one am marveling at how there have not been more. I actually have a theory that the natural warming we have seen over the last 30 to 40 years could impede tropical cyclones by changing the large-scale pressure and wind patterns in the Northern Hemisphere to disturb the mechanisms needed for the severe seasons. But that is another story, one best reserved for weather geeks like me, which is where the whole CO_2 debate should be. It should be a discussion in search of an answer, not something being pushed as a doomsday machine on a nation facing far bigger challenges far quicker.

Of course, the Paris climate accord was a big deal, and President Trump's pulling out raised the ire of the man-induced climate change propagandists. (I refer to them by different names because they have so many different ways of referring to themselves. Sort of like claiming anything that happens is because of what they are saying.)

I wrote "More Paris Madness" on June 9, 2017. The Associated Press had written this headline: "Climate decision could accelerate damage to Trump properties in Florida." So I responded:

> This is part of a slew of doomsday proclamations that have increased since the Paris situation. But this is another good example of why *this is about economic considerations.* Because something as bad as what happened in the 1940s — a railroad track of major hits in south Florida — is bound to happen again naturally.
>
> Even if Donald Trump had stuck with the Paris climate accord, it doesn't mean a hill of beans as far as the threat to our coasts. And given factual past events, one would expect that.
>
> The reason so many people are in harm's way is because, overall, recent hurricanes have not been that bad, so people feel the risk is worth it.
>
> In my hurricane forecast this year I explain my idea on why we haven't had the big powerhouses maintain intensity as they barrel toward the coast, which of course means build, build, build, because you thumb your nose at nature only to start looking for something (or someone) to blame when she does what she naturally does. But this year looks different, and so the ignorance on display in the articles like the one above — and they are flying off the presses left and right since Trump chose to abandon the Paris climate accord — is startling.
>
> I am sure the author did not read my hurricane write-up because if that person did, he or she would have to conclude that all the resorts near

Mar-a-Lago are under threat of being devastated. It has nothing to do with signing a non-binding accord that economically is all pain and no gain for the United States. Trump wants to build in Florida, whether he is president or not. He shares the same risk as anyone else. It's part of a longstanding pattern that, if looked at, would show the inherent risk.

As you can see by what I said above, 2017 was no surprise even with the climate alarmists pushing it as part of their agenda. The fact is that CO_2 had nothing to do with the rise of the high-impact season; it was a pattern that we envisioned before the season started based on past patterns that looked similar.

When Al Gore won the Nobel Peace Prize in 2007, I was incensed for three reasons. 1) It was based on ideas that in any other non-agenda-driven field would have been laughable. 2) It was partly because of a warning of future events. 3) He beat out Irena Sendler, who actually did something heroic in arguably one of the darkest examples of man's inhumanity, the Holocaust. Her actions in the Warsaw Ghetto in saving 2,500 Jewish orphans lost out to Al Gore?

When Al Gore's sequel came out, it gave me another chance to bring up this example of true heroism.

A lot of people got mad and accused me of cherry-picking when I pointed out that 2017, even after the major Super Niño of 2015-2016, started off colder than 2007, the year Gore won the prize. The fact is that temperatures overall have not moved much since 2006, and if one incorporated 2012 — five years after Gore was awarded the Nobel Prize — it would have been colder overall than the run-up to Gore's first movie, but the Super Niño brought it back up. But January 2007 had a global temperature of .555°C above the 30-year mean, while 2017 was .411°C above that average.

On Aug. 8, 2017, I wrote "The Value of Al Gore's Movie: Rekindling the Bravery of Irena Sendler."

I liked "An Inconvenient Truth" right up there with "The Wizard of Oz" for Wild Weather Fantasy representation. But with the advent of Al Gore's new movie, it brought back a much more serious travesty: his winning of the Nobel Peace Prize over Irena Sendler.

A lot of people may not know who Irena Sendler is. For those who don't you should google her name and see what she is about.

There is a caution at the end to disregard a lot of the "Snopes" comments. Interestingly enough, Snopes seems to be imploding.

There are some movies about Irena Sendler, like "Irena Sendler: In the Name of Their Mothers" and "The Courageous Heart of Irena Sendler."

"An Inconvenient Truth" is well known. These are not. While Irena has won some awards, Al Gore's new movie reminded me of what a travesty it was for him to even accept the Nobel Prize over what this lady did.

Just what did she do? From irenasendler.org: "Irena Sendlerowa was a Polish woman who, along with her underground network, rescued 2,500 Jewish children in Poland during World War II. Many of this number were already outside of the Ghetto and in hiding."

Side note: Most of the children's parents died at the Treblinka extermination camp.

She died in 2008.

This is fascinating to me on many levels.

Her deeds are accomplished fact. They happened. They took a degree of courage that only one who actually faced such events could understand. I know I can't.

Al Gore was awarded the Nobel Peace Prize based on warnings of future events—the same future events that have not happened. The fact is that global temperatures from January 2007 while Gore was basking in the glory of his apocalypse-driven fame *were warmer* than they were in Jan 2017, and we are still falling off the Super El Niño peak. Additionally, much of the time in-between was lower than what it was in the run-up to "An Inconvenient Truth."

But there is more to me. Let me lay my cards on the table. Over the years I have become a big fan of Israel. I am not Jewish, but I find the history of the Jewish people remarkable, if not astonishing. What happened in World War II cannot be put into words. Here we have a case of someone with actions far beyond the fantasy of a forecast that took a back seat to ... what? Now let me ask you this: If you were in the running for the Nobel Prize against Irena, would you even accept the award understanding that what you are doing involves an agenda that is relying on future events versus actual heroic accomplishments in one of mankind's darkest hours? Who would do that?

Even more distressing is the idea that you actually equate your cause with causes that have real value for the people who are involved in them. For instance, equating "climate change" with racial equality. That is a flat-out insult to that cause. Or labeling people who disagree with you as "deniers" or "Nazis," which shows total disrespect for people who can never forget what happened in one of mankind's darkest hours. Shame on you. Shame on you trying to equate your straw man argument with real problems that people bled and died for and the problems the world faces today. Shame on the people who think that the heroism of the past is less worthy than fantasy-driven utopian agendas of the future.

I am glad Al Gore has his new movie out. It reminded me of Irena Sendler, whom he beat out for the Nobel Prize. Because it gave me a chance to write on someone whose story should be known and once again expose someone who has gotten rich off something that can't hold a candle to the bravery of people in the era that Irena Sendler exemplified.

What the politicization of weather and climate means is that there is no possibility for an objective argument in the true sense of science. There is simply too much at stake. Therefore, we see a march toward the shutdown of free debate. And above that, we see another layer of mud being thrown on something I love. What am I to do? Try my best to wipe off the mud. It comes down to this: If it wasn't political, would it be important? Interestingly enough, because climate change has become important for the wrong reasons, and it's simply a means to an end, it means it's virtually impossible to get the right answer. Like in so many things in politics, it's caked in a layer of mud.

Chapter 5

Academic Mayhem

The words of Eisenhower ring so true today when it comes to the role of academia in facilitating agenda-driven issues, and the climate is one of them. Eisenhower said:

"The prospect of domination of the nation's scholars by federal employment, project allocation, and the power of money is ever present and is gravely to be regarded. Yet in holding scientific discovery in respect, as we should, we must also be alert to the equal and opposite danger that public policy could itself become the captive of a scientific-technological elite."

The link between a larger federal government and academia, which by and large believe in top-down authoritarian control while parading in the guise of the free exchange of ideas, has no better vehicle than the climate debate.

Yet in knowing how hard it is to earn a PhD, I have respect for people in meteorology simply because they have done it. But I think there is something more than that driving all of this, and it has to do with what I think is almost a revenge of the nerds. At Penn State, for instance, there was always a jock-against-student attitude. Being that I was both (not very good at either, but I did wrestle and I did meet the requirements for my degree), I hung out with both sides. Football was king at Penn State and brought in an immense amount of money, but I always sensed a lot of the serious teachers were very jealous of that. Now, especially in climate, just as many people know the major climate players as they do the football coach. I am not begrudging either; it's just that while I knew who Bill Koll (my wrestling coach) and Joe Paterno were when I was in college, no one knew who the people running the PSU meteorology department were, yet they were like gods to me (Al Blackadar, Charlie Hosler, John Cahir, Joel Myers, etc.). Today there is a great deal of fame that comes with being a climate rock star. Interestingly enough, if James Franklin were losing football games, he would not be there very long. Every week during the season, James Franklin's football team gets "questioned," and if it's wrong enough (losing games), James Franklin knows he is gone. In the climate situation, not only is there no admission of forecasts that have gone awry, but we are not even allowed to question them. It would be like a football season with no game.

In 2013 I penned this: "Can an AGW Climatologist Be Truly Objective?"

I am not someone who demonizes Dr. Michael Mann. Because I have read almost everything he has written and see exactly where he is coming from, the totality of the journey is something I look at. In fact, in some public speaking engagements, people get upset when I tell them to go look at all the work he has done. As I said, I am an issue-by-issue person, but in penning this piece I realized how much someone like him has at stake here. For me, the weather corrects me every day, and I have to deal with the fact that patterns are warmer no matter what the cause. I just don't believe CO_2 is the climate control knob. But imagine if everything you are known for by the large majority of people was proven to be wrong. What happens? Add in the idea that you are fighting to save mankind and, well, it's pretty hard to turn away from that.

The means have to justify the ends. Too much is at stake and there can be no turning back.

I would love to debate Dr. Michael Mann. He's a professor at Pennsylvania State University, and I'm a Penn State grad (Meteorology, 1978). Enough people know me, as well as him, so we could charge a modest admission, fill Eisenhower Auditorium at PSU, and give all the money back to the PSU meteorology department that I still love dearly in spite of my outcast status on the anthropogenic global warming issue.

But Dr. Mann would probably want no part of debating me on the main drivers of weather and climate given I have no higher degrees. C'mon, a BS in meteorology from PSU against this (via Wikipedia)?

Education: A.B. applied mathematics and physics (1989), MS physics (1991), MPhil physics (1991), MPhil geology (1993), PhD geology & geophysics (1998).

Alma mater: University of California, Berkeley, Yale University.

This would be a blowout. What chance would I have?

Let me be clear: Dr. Mann's resume, and those of anyone who receives a PhD in the physical sciences, impress me. I've read almost everything Dr. Mann has written and, because of that, I understand where he's coming from. But there are things he's lacking when it comes to the pursuit of the right answer, and that's the methodology one learns in putting together a forecast. One must examine all of what his opponent has, not close his eyes to anything that might challenge his ideas.

For instance, while I've read almost everything Dr. Mann has written, how many times has he had hands-on experience in making a forecast that has to verify? It's laughable to think as a private sector meteorologist whose livelihood depends on being right that one can separate climate from weather. I realized a long time ago that being able to recognize current patterns from understanding the past (it was drilled into me by my father, a degreed meteorologist) was essential to making a good forecast. The fact that many climatologists downplay the relationship, or say they're different, shows me they don't know what they're talking about. In other words, I do what they do, but they don't do what I do. I read what they write, but they won't stop to look at the other side.

Perhaps it's like something we sometimes see in sports — the curse of talent. Most of these people are *very* smart. I went to school with future PhDs and could see that in the classroom, they were like my wrestling coaches at PSU — guys who were great doing what came natural to them. However, my wrestling coach used to stress that when you're used to having everything come to you, it's very hard to change and step up your game. Consequently, you'll get beat on your weakest point and what you don't know, and that's where the methodology in forecasting comes into the climate debate.

You see, in what I do, one must weigh factors and decide which ones are most important. Additionally, one gets used to challenges that can never really be seen in research. How so? Suppose someone gives you a grant to study global warming. Can you come back and say, "My research says there's no global warming"? You have been given a grant to produce a result; how can you possibly justify that result if it's the result that would cost nothing to come up with in the first place?

In my line of work, getting paid (having clients) *depends* on the correct result. The client doesn't say, "I want a cold winter, here's the money, forecast it." The client asks for a forecast that gives him an edge. If you are right, the client renews; if not, it's bye-bye. But there's no upfront money that looks for a set result. This means the forecaster does not care whether it's warm or cold, just that he gets the right answer, *whatever* that may be. This is not the case in the AGW branch of academia. Research grants come with the cause du jour — just try getting a grant to disprove global warming (actually, you don't need one; it's easy to refute it just by understanding what's happened before).

That said, regarding the climate debate, what factors am I looking at to come up with my conclusion? To me, this is a big forecast, and the simple answer is: It's hard to fathom that CO_2 can cause anything beyond its assigned "boxed in" value with regards to temperatures because of all that's around it. It comes down to the sun, the oceans and stochastic events over a long period of time — action and reaction versus a compound comprising .04 percent of the atmosphere and 1/100th of greenhouse gasses.

But unless you work every day in a situation where you are reminded you can be wrong, you don't have appreciation for the methodology of challenge and response you need to be right!

Then there's another big problem: What if you have all this knowledge, you've taken a stand on this, and it's your whole life — how can you possibly be objective? The climate debate and past weather events are needed building blocks for my product. That product involves a challenge each day. In the case of PhDs on the AGW side, they believe the idea *is* the product. Destroy the idea, you destroy the product; destroy the product, you destroy the person. Therefore, it's personal. Your whole life — all the fawning students, the rock star status — is all gone. I would hate to be in that position. Each day I get up, there it is — the weather challenging me. The answer *is the fruit* of my labor, not the *object* of it. Because of that, you'll look for anything to come up with the *correct answer*, not just a predetermined one where your self-esteem depends on it.

So, these giants of science have a fundamental problem, and it runs contrary to their nature. In the end, the very talent and brilliance of a lot of these people may be what blinds them to what it takes to truly pursue the truth.

On July 10, 2015, I wrote "An Open Question to Our Universities on AGW: What if You Are Wrong?"

No climate program in the nation that I can find is teaching anything but the "party line" on global warming. Perhaps they are right, though if the forecasts made by computer modeling were graded like any college exam (if you said one thing was the answer and it turned out it was wrong), then I would expect the professor to grade it as being wrong. I assume college courses are still graded based on correct answers.

In my world it comes down to this: The trends can easily be seen — the warming that was shown *by all* measuring tools through the late '90s and the recent downturn or even slight cooling over the past 10 years in the model analysis.

Of particular interest are the past two El Niños that had "spikes" in '07 and '10. But the downturns that followed wound up offsetting the spike. Also of interest is this question: When it comes to model analysis of global temperatures, why would the NCEP come up with a global temperature that does not support the idea that the earth is still currently warming, as a recent NOAA press release implied? Here is why the coming five years are huge: If, as I believe, the coming El Niño spike is followed by a bigger downturn, will the universities, lined up solidly in the anthropogenic global warming camp, admit there is a problem? The current El Niño excitement is similar to that of the great El Niño of 1997-1998. Many on the AGW side are opining this will lead to a "step up" to a new plateau. And they certainly have a point here, in my opinion. If there is a "readjustment" up, then it will silence me for one. In other words, if the Super Niño simply establishes a higher base point, it still won't mean that CO_2 is the cause; it would mean that my ideas on the cooling were wrong.

But I am curious, given the constant drumbeat that comes from government, media, academia and, most recently, papal authority and immense investments: Is there any way out for the other side that will not completely ruin them?

There is so much behind this movement, I really can't see any way the people driving it can possibly back away. As for me, it's simple. This El Niño and the three years that follow with objective global temperature recordings will answer it for me.

In a world spinning out of control, we are asked to believe that global warming is the biggest threat to mankind. I ask people of goodwill who don't see things my way to ask themselves this question: What would it take for you to at least have doubt (that should be natural in any future event) and to change your position? I even wrote on that: "Is There Anything in the Global Warming Debate That Would Convince Me I'm Wrong?"

Perhaps there is hope. Dr. Ivar Giaever, a real Nobel Peace Prize-winning physicist and former professor at the School of Engineering and School of Science at the Rensselaer Polytechnic Institute (and also a supporter of President Obama), said this: "I would say that basically global warming is a non-problem." He added, "I say this to Obama: Excuse me, Mr. President, but you're wrong. Dead wrong."

To know for certain one is right, one has to also know for certain what would prove himself wrong. Only by that kind of open-mindedness can one really search for the truth.

I am not in the habit of attacking academia, but I do believe that politics and ideology have taken over. The universities were always supposed to be the bastion of open-minded debate, but that is not the case here. I understand why — livelihoods are on the lines, egos are on the line, agendas are on the line, and, yes, the answer to the question is on the line. Only one of these is focused on the last part. The other things severely limit that chance, if not reduce it to zero. Sad to say — and I have only shown a couple of examples to make my point — but the education system is not advancing this but acting in a repressive manner to stifle the very questions needed to come up with the answer.

Chapter 6

Odd Things as the Means to Justify the End

There are two sets of people in the anthropogenic global warming debate. First, there are the serious scientists who have great points on the matter. Interestingly, whenever I run into them, the talk is always cordial. And given that they can tell me what they are looking at, I enjoy talking to them because they give me things to look at that challenge me. In fact, I wrote a blog on that earlier in the book.

But then there is the group that cannot even be spoken too because their mission is making sure that no matter what happens, their answer is the only right answer. When I say AGW Ilk, Climatariat (Judith Curry's term), Warmingistas, Parasitic Climate Ambulance Chasers, etc., I am both having a little fun and simply giving back what they throw at me, although those terms are not nearly as demeaning as calling someone a denier of something he does not deny in an effort to smear their reputation and also link them with a period in mankind's history that is as despicable as they come.

The problem in this debate is that if you turn the other cheek, they hit you again. Open hands are met with closed fists.

Most of the latter are politicos, actors, activists, leftists, etc. Basically anyone who believes the means justify the ends.

If we're paying attention, we find that climate alarmists constantly shoot themselves in the foot. It's like they are unaware that what they say makes little sense if the public stops to think about it.

"All This for .01°C?" was written in August 2015 and is a classic.

When I heard former EPA Administrator Gina McCarthy make an absolutely absurd comment regarding the Obama administration's climate accord agenda, I thought she was purposely trying to stop Obama's own agenda! In a nutshell, she was saying that all the steps we were taking that would negatively impact our economy would save .01°C of warming — something that means next to nothing (if even true) for the climate and can't even be measured accurately because it's within the margin of error. It was being done to be an example for the rest of the world — the U.S. would lay down its economic lifeline and redistribute its wealth so everyone else would be happy and follow along. It was truly one of the most deer-in-the-headlights moments of the entire debate that went largely unnoticed. I could not believe it.

From the blog:

> As the president reveals his plan to reduce greenhouse gases to save us from an apocalyptic atmosphere, I wish to remind people of three things:
>
> 1) The true hockey stick of the fossil fuel era — global progress in total population, personal wealth and life expectancy — shows that, because of fossil fuels, world progress is heading up in a hockey stick-like fashion.
>
> This is truly amazing. To show how fossil fuels played a role in expanding the global pie, there are many more people alive today who are living longer and enjoying higher GDP. One has to wonder if someone against fossil fuels is simply anti-progress. Which is ironic since many climate alarmists like to label

themselves "progressive." They're certainly anti-statistic given this inconvenient hockey stick staring them in the face.

2) The geological timescale of temperatures versus CO_2 shows no linkage. As much as I struggle, I can't see the linkage.

3) The EPA's Gina McCarthy admitted that the steps being taken would only prevent .01°C of warming, but it was the example that counted for the rest of the world.

This in addition to the fact that in 2011 she admitted she did not know how much CO_2 was in the atmosphere. And the EPA's lines of evidence for this *are provably false!*

Given the facts, I can't help but wonder: Did policymakers ever take Economics 101 or even a course on how to read a chart?

When I see simple questions that raise doubts or even outright debunk all this, it's like watching the opening from the old "Twilight Zone" series: "You are traveling through another dimension, a dimension not only of sight and sound but of mind. A journey into a wondrous land of imagination. Next stop, the Twilight Zone!"

Before writing this piece, there were several other pieces that showed how convoluted this all is. The above piece was a big whammy, and it revealed an amazing ignorance as to what was going on. If it was not ignorance, then the assumption had to be that the public is too stupid to realize the absurdity of the U.S. sacrifice over nothing. Or perhaps the administration simply assumed no one would pick up on it, which, sadly, was true in many cases.

In any case, I wrote "Data Deception" on Aug. 16, 2012, when the idea that July 2012 was the hottest ever was being pushed. (Interestingly enough, we had cooled globally to near or below the 30-year running mean.)

The recent announcement by NOAA that July was the hottest ever has drawn counterarguments that prove it was not.

It is true that it is very warm off the Northeast coast, which is part of the reason I am so concerned that, like the 1950s, the Northeast is vulnerable to hurricane landfalls. However, it's very cold off the West Coast. That should be relevant also when talking about how warm the waters are off our coasts. It's part of the cold cycle of the Pacific we have entered.

I wish to make a few points. There seems to be a common thread with people who believe their idea of absolute truth means they can say and do whatever they want to get to that truth. The result is a series of loud announcements that grab the attention of a gullible media but when debunked do not have near the coverage as the original pronouncement.

So here are some points. Though attributed to Joseph Stalin, there are actually hosts of people who may have said this, however, its meaning is clear: For the greater goal, there must be casualties. In the case of what we are confronting, belief in the end result means that there has to be suffering, be it the truth, people, individual freedoms, or the economy. The bullhorn screams doom, the public hears it, and the truth is not heard.

- A significant quantity of weather stations have been compromised by the urban heat island effect. The station data used in the 1930s, when the heat was just as bad or worse, was not nearly as contaminated as it is today. If we remove some of the weather stations that are skewed by urbanization, the heat is not as bad as the 1930s. There are some important points that need to be separated. One might argue that heat islands should not be overlooked because they indeed make the U.S. and planet warmer. However, the amount of land covered by cities pales in comparison to the amount of land/ocean that is not covered by cities. Therefore, the overall effect of heat islands on temperatures is miniscule. However, because a disproportionate amount of weather stations is situated in cities, and thus susceptible to the heat island effect, the calculations used by global warming activists are severely biased. An accurate calculation of global temperatures requires the removal of weather station bias. Additionally, the heat island effect has nothing to do with carbon dioxide emissions and the draconian energy policies that are destroying our economy. And if it was comparatively hot in the 1930s (and in the 1950s), why is CO_2 the cause now but not then? The conclusions of climate alarmists about what this means are disturbing to any rational person.

- The U.S. is about 2 percent of the earth's surface. For a movement that is so concerned about the planet, why is it ignoring the rest of the planet? The Southern Hemisphere is having a very cold winter while the Northern Hemisphere had a cold winter last year, despite North America's warmth.

- "You can't make an omelet without breaking a few eggs."

It is either ignorance or deception, probably both. What else can we say? Are they ignorant of the facts? Do they even look at the other side of the argument? Is their confidence built on ignorance of the other side of the argument? (The Dunning-Kruger effect). Or do they know the facts and simply put out deceptive statements meant to confuse the public, forcing them to rely on the state to "save" them and the planet? There is only one way to fight this. We must constantly challenge the public to look at all sides of the argument so they can understand what is going on. This is an uphill fight, but persistence overcomes resistance.

I will end with this popular quote, one that is familiar to most of us:

"The only thing necessary for the triumph of evil is for good men to do nothing."

Doing nothing in this matter is not an option for those concerned about the truth and the ideals our nation was built on.

Here we are five years later and the ideals our nation was built on are being roundly attacked. And there is the point — it's not about climate and weather. It never has been with a lot of the people involved here.

The Climate Chronicles

On Feb. 14, 2013, I wrote "Deceit or Ignorance? You Be the Judge."

The recent blizzard in the Northeast was called first by Weatherbell.com on Feb. 3, the Sunday before the storm. I pulled out maps of the infamous Lindsay Storm of Feb. 9-10, 1969, and then compared this storm to it — *five* days in advance. At that time, other forecasts for Friday in the blizzard impact areas were for highs between 40 and 45 degrees with a *CHANCE* of rain or snow.

A forecast calling for a chance of rain or snow and a high between 40 and 45 degrees does not describe what happened, just as the forecasts calling for Hurricane Sandy to escape out to sea as a subtropical storm don't either. Because of this, the "unexpected" nature of the storm that the public was led to believe in supplies fertile soil for those who drive the AWG propaganda machine and put forth misinformation about these storms.

Unlike the Lindsay Storm, where the epicenter of the 15-30 inches of snow was almost right over New York City, this was a bit further east. The point is that, once again, in spite of a predictable and similar event to what had occurred before, this is being used as a propaganda tool for global warming/climate change folks (I keep getting the terms mixed up — it's bound to change tomorrow). They're either simply lying, or they are relying on a negligent low-information population so they can then try to convince them that everything that happens is because they are right.

Why should anyone trust people who had no idea five days before what was going on when they come out after the fact claiming it's because of global warming/climate change? (Want to make sure I have the right term covered…)

Now let's get to Sandy.

I think I am qualified to talk on Sandy given I predicted the hit *nine* days beforehand to both clients and the public for the area impacted.

In September 1967, Doria, a Category 2 hurricane, backed into the East Coast. Had it started its westward run at a slightly more western longitude, Doria would have devastated the East Coast as bad or worse than Sandy.

Again, it's a matter of a hundred miles, *nothing* within the realm of what the atmosphere can do.

But let's take a look at several more storms:

- *Category 4* Hurricane Hazel in 1954, which hit the North Carolina in mid-October and whose upper pattern I used to set up Sandy for her run.

The track of this monster: A north-northwest path into the Carolinas, then almost due north afterward — in October! Hurricane force winds all the way to Toronto!

What if the track was 150 miles east of that? A Category *4* hurricane would have hit the Mid

Atlantic coast running in from the ocean. Again, that kind of track difference is nothing in terms of the weather, a whim of the pattern at the given time.

• The 1938 hurricane.

The track did not veer out to sea. Instead it came north through Long Island with reports of a 30- to 40-foot storm surge and wind gusts to 186 m.p.h. Five-minute sustained winds registered at 121 m.p.h. at Blue Hill, Massachusetts! It's hard to even comprehend that!

Seventy-five miles further west and the full weight of the storm surge is into New York City! In fact, the disaster scenario played out in the storm surge models has a 20-foot-plus scenario in New York City, and this has been talked about for over 25 years as something that we may have to contend with one day — not because of climate change or AGW (whichever it is) but because that is what *nature can* and probably one day *will do*!

When you look at these tracks, it's simply the whim of the weather that a storm *worse* than Sandy has *not* happened. There is nothing magical or mystical about this.

Or how about the 1903 hurricane in Atlantic City that hit from the southeast?

Or Agnes in 1972 with its devastating floods and *snow in West Virginia* — on the first day of summer!?

What you don't know *can* hurt you, as we see people who are either ignorant or *know the facts and are trying to deceive* people so they can push their agenda. It has to be one or the other. You either don't know, in which case you are not qualified to make such statements, or you do and are trying to deceive people.

So is it deceit or ignorance or just one honest mistake after another? You make the call!

On April 8, 2013, I wrote "Forecast: No End in Sight to AGW Excuses."

You can't make this stuff up. The climate change "authority" has published several articles on how global warming is resulting in an increase in Southern Hemispheric ice, while the northern ice cap is supposedly set to disappear by 2050 — even as global sea ice is back above normal. I do not believe it's a coincidence that the articles are being released as the Northern Hemisphere fights its way out of what seems to be a never-ending winter in some places. Here in the States, the Plains are having the opposite weather of last spring, with record cold and snow so late in the season. Last spring was touted as an example of global warming with the record warm March here in the U.S., even though the *globe was .106°C below normal!*

Let's examine this one at a time.

After years of stories on the shrinking of Antarctica we find in an article by the BBC that the expansion of the ice cap to near record levels is because of melting underneath, leading to fresher water near the surface that freezes quicker. I have a few questions, given the actual Antarctic ice cap levels at the time were above normal.

- Why wasn't it explained to us *before* why this would happen?

- How were we measuring temperatures at all these deep levels 40, 50 or 60 years ago?

- What kind of historical (hysterical?) data do we have to compare this to?

When blaming man for all the increase in warming (a joke in itself), here is something to consider. An article by Neil Catto, titled "Revisiting Temperature Reconstructions used in Climate Change Modeling," deals with temperatures in the UK. It includes a great passage on how much CO_2 in the atmosphere the UK is responsible for:

"CO_2 levels at present are near 390 parts per million of the atmosphere (0.039%), which includes water vapour and other GHGs. The natural carbon cycle produces 2960000m tonnes CO2. Mankind's contribution is understood to be 33500m tonnes which equates to 1.13%. The UK contribution of 458.6m tonnes equates to 0.0155%. Therefore the total atmosphere (all GHG) is 2960•0.039%•100=758974358974m tonnes.

"So, the UK's CO2 percentage of global atmospheric gases is: 0.0000000604%."

The U.S. is responsible for 10 times the UK production, so the U.S. CO_2 percentage of global atmospheric gases is .000000604 percent.

If you really want to have some fun with numbers, consider this: The oceans have 1,000 times the heat capacity of air. Assuming air is a homogeneous mixture, with CO_2 being 1/2500th of air, it would mean that in the ocean/air system, CO_2's contribution is 1/250000. According to the article above, man's contribution to CO_2 is 1.13 percent. I have seen figures as high as 5 percent, so I will use 5 percent as the baseline. Man's contribution to the heat capacity of the entire system would be 1/5,000,000. Since the U.S. is responsible for 20 percent of that figure, our contribution to all this is 1/10,000,000.

So over this we should enact draconian laws that slow down our economy and inhibit the chance for people to prosper?

And we are expected to believe that this is causing the earth's climate to change?

For the record, I am *not* anti-alternative energy. In fact, I am quite the opposite: If cooling occurs to the extent many of us think, we are going to need all the energy we can get our hands on. It takes more energy to heat cold homes and businesses than it does to cool warm ones. Life does better when it's warmer, not colder. I am not for shutting down research and development. I am for identifying problems and having the freedom to confront them with sound, rational approaches based on reality, not excuses.

Keep in mind, the rise of CO_2 yearly is a bit over 1.5 parts *per million*, and in charts CO_2 is given a disproportionately large look. It's really a very minute rise compared to the fluctuation in the ocean and air temperatures. The point though is that there is a CO_2 disconnect, with ocean and air temperatures moving in tandem.

Let's go to the Northern Hemisphere.

Here we are with one of the coldest March-April back-to-back periods in 50 years in the Northern Plains, and we see that the northern ice cap is now forecast by NOAA to disappear by 2050.

Amazing how these articles come out when major cold is blasting the U.S.

Why would anyone believe a NOAA study that says by 2050 the northern ice cap is gone when natural climate theory allows for the back and forth of the ice caps? In fact, the expansion of the southern ice cap is enough that *global sea ice is above normal.*

It's a darn good thing we don't have the north above normal too! In any case, check out the article, because if you are still wondering where spring is, you can take comfort in the prediction of an ice-free Arctic by 2050. (Don't you love these forecasts that no one will be responsible for when their verification time arrives?)

Remember last March how it was offered up as "proof" of global warming? The reality is that March *globally* was *below normal* (-.106°C).

The excuses reach full stride when one sees that the *increase* in Southern Hemisphere ice at 4 percent per decade is outpacing the *decrease* in the Northern Hemisphere at 2.5 percent per decade.

Common sense tells you that this is cyclical; that such an imbalance will self-correct when the Atlantic warm cycle shifts. Global ice is *greater* than it was.

Much like I explained in a **previous article** that exposed the AGW double standard, the climate change propagandists will claim they are correct either way: The decrease in ice in one place is because of warming, the increase in another also because of warming. But how is it that someone like me — one who doesn't believe in any of these things — can forecast long term and people who can get a forecast for free actually pay me? After all, without the knowledge of what CO_2 is doing, how can I have any credibility?

Again, let's look across the pond where there's more "down is up, up is down" excuse-making. This time people in the UK are being asked to believe that **warming is causing them to be colder** — even though less than 10 years ago their same forecasters were claiming snow and cold would be a thing of the past!

I am involved in UK forecasting and we had a very cold winter forecasted. In fact, an article by Matt Ridley in the Spectator Diary shows that:

"At least somebody's happy about the cold. Gary Lydiate runs one of Northumberland's export success stories, Kilfrost, which manufactures 60 per cent of Europe's and a big chunk of the world's aircraft de-icing fluid, so he puts his money where his mouth is, deciding how much fluid to send to various airports each winter. Back in January, when I bumped into him in a restaurant, he was beaming: 'Joe says this cold weather's going to last three months,' he said. Joe is Joe Bastardi, a private weather forecaster, who does not let global warming cloud his judgment. Based on jetstreams, el Niños and ocean oscillations, Bastardi said the winter of 2011-12 would be cold only in eastern Europe, which it was, but the winter of 2012-13 would be cold in western Europe too, which it was. He's now predicting 'warming by mid month' of April for the UK."

(By the way, it is warming there as the forecast said for mid-month.)

I have made this point many times. We are dealing with people who make bombastic statements. And when these bombastic statements don't verify, the charlatans claim that what happened is *still* because of what they wrongly said. It would be ripe material for an SNL comedy skit, except the harm this non-factual agenda is causing to the advancement of mankind isn't humorous. Even as I write this, I hear that Maryland is adopting a "rain tax" for objects that prevent water from flowing naturally into the ground. Actually, that's closer to something that does have an effect on temperature — water vapor — so in a way Maryland is finding a way to tax water vapor, or a byproduct of it.

As the cold in the U.S. reaches record levels this late in the season, do you notice what is going on? There are articles about melting ice or ice expansion *because* of warming, but nothing about how cold it is so late compared to last year, with the drumbeat of how warm it was so early (even though the globe was colder than normal). It's obvious to anyone who possesses common sense that this is agenda-driven. And while the actual forecast for tomorrow, a week, a month, or years from now has some degree of doubt, there is one forecast we can be sure of: There is no end in sight to the "climate change" excuses.

The purpose of this chapter is to show what climate alarmists are up to. And there is a lot to show (my examples are only the tip of the iceberg, which by the way do still exist).

On Oct. 12, 2013, I wrote "Anything Can Happen and Probably Will." The climate alarmist agenda has numerous people funded to keep up the drumbeat, and I get mail all the time showing me what is going on. So here is what I wrote:

I am going to do something a bit different here for *Patriot Post* readers. With the upcoming winter looming and the climate alarmists lying in wait for anything that happens anywhere as evidence to prove their point, I thought it would be nice to give an example of how no matter what happens, it will be repackaged as proof humans are wrecking the climate. It seems like a new strategy has evolved — using people who really don't know the weather and climate, repackaging knowns and then claiming it is some big discovery that backs their idea. I used an example last week: The "hidden heat" in the ocean, which Dr. Bill Gray explained over 30 years ago with his ideas, forecasting the current overall weather pattern that lead to the increase in hurricane activity.

I caught a tweet from Bill McKibben, the head of 350.org who has a BA from Harvard, but it does not appear to be in meteorology or climatology.

I want to be as fair as possible, and nothing can be fairer than your looking at exactly what this organization and the people who make up its team are about.)

Dr. McKibben apparently ignored, or did not know, about the forecast I made a week before when I asserted that what I was most sure of is that climate alarmists would use the tropical storm and snowstorm occurring simultaneously as an example of ... whatever it is they are pushing. More predictable than the weather is the prostitution of weather events by these people — events those of us who have loved and studied weather and climate all our lives know about. The late Dr. William Gray said it best and was most to the point. He equated the taking of grant money for the sole purpose of pushing this agenda as akin to being a climate whore. Dr. Gray never was one to mince words.

Let's look at McKibben's idea.

The snowstorm was a record-breaking early season event associated with abnormally cold air. For it to snow that much this early, we had to have *well* below normal temperatures. It occurred in an area of the nation where late-season cold was delivering snow into May. So let me get this straight: Late-season cold followed by early season cold is somehow indicative of the whole global warming disaster?

And what about Tropical Storm Karen? Apparently, Dr. McKibben was not aware of the reason Karen was unusual. It is very rare for a tropical cyclone in the Gulf of Mexico to *not* make landfall as a tropical cyclone. In other words, what happened with Karen is evidence *against* what he is trying to push. *The storm died in the Gulf,* though its ghost is delivering more wind and rain to the Mid-Atlantic states now than it ever did in areas of the Gulf Coast under a hurricane watch. But facts mean little to these people. Grab the headline and never mind the truth.

(Side note: The propaganda about the latest F5 tornado ever in May so far south, which was downgraded to an F3, failed to mention the cause of it — the result of a major cold trough abnormally far south. What was remarkable about it, had it been an F5, is the fact that it was so late and so far south because it was so cold.)

We had a tornado outbreak in the Plains last week. I guess Dr. McKibben is unaware of the second season that occurs with tornadoes. As the upper jet starts to intensify in the fall, and the upper air temperatures fall, warm humid air masses can come in off the Gulf of Mexico. The result is a spike in the number of tornadoes in October and November. In fact, the second season is particularly dangerous because often times these tornadoes occur at night. Houston, Texas, was hit on Nov. 16, 1993, in front of the nasty winter of '93-'94 across the U.S.; Huntsville, Alabama, on Nov. 15, 1989 (the following December was one of the coldest on record, by the way); and Shreveport, Louisiana, on Dec. 3rd, 1978, again in front of a major cold winter. Interestingly, if I were him, given the *lack of tornadoes* this year, I would be worried that the occurrence of twisters late in the season is a harbinger of a cold winter. But then again, I am sure a cold winter would simply be twisted into more evidence that we are heading for a CO_2-fueled climate disaster.

And then there are the wildfires. We have a near record low year, yet the wildfires that are showing up are used as evidence of the weather going wild? Again, never mind we are so far below normal that the opposite is more of an option to a rational person.

Is there anything truly rational about a group which thinks that no matter what happens, it shows it was right, even if the results are contradictory? You make the call on that.

This is a classic take-credit-no-matter-what example, one that I can shoot full of holes with past events. I am sure Bill McKibben is a very nice man. Having read his background, I am sure he and I share a lot in common in spiritual matters. Yet he represents people who are so heavily invested in this that they cannot see the other side. I point this out time and time again. The global warmth is dealt with from a forecasting point of view no matter what the cause. In relation to what I do for a living, so what? It's warmer, therefore you must deal with it. *But if the cause and desired outcome from the other side are taken away, so is their life's work.* I do not envy anyone on the other side of this issue, for they have no choice in the matter.

As the chapter title says, it leads to odd things as the means to justify the end. It is such that "Sometimes You've Just Got to Laugh," as I wrote on Feb 14, 2014.

> I don't watch the Olympics that much (except for wrestling, which I think should be moved to a winter sport since it would give it more visibility), but I used to love watching figure skating. However, the SNL skit with Chris Farley ruined it for me.

> Even the Olympics is now being turned into a propaganda message, as we are hearing about how warm it is. Um, hello? Sochi's normal high is 52°F in winter.

> Here is what Wikipedia has to say about Sochi:

> "According to the 2010 Census, the city had a permanent population of 343,334, up from 328,809 recorded in the 2002 Census, making it Russia's largest resort city. It is one of the

very few places in Russia with a subtropical climate, with warm to hot summers and mild winters."

You are holding the Olympics in a subtropical climate and you are complaining about it being warm?

Or how about the reason it's so cold in the U.S.? I think Arctic ice melt was one of the reasons. The winters of the 1970s were comparably cold. The winter of '76-'77 was one of the warmest on record in Alaska and much of the Arctic, with near record *high* Arctic sea ice. So why is it at that time, at the height of the ice age scare, it wasn't the same thing as being pushed now? Because there is no difference — it's natural and cyclical, and since we came out before the fact with a cold and stormy winter, modeling it after 1917-1918 and 1993-1994, obviously it's something that has happened before.

By the way, the fact is, during this current cold period here, the earth has been averaging *below normal* temperatures as the herky-jerky downturn continues. The globe has been averaging between .2° and .3°C *below* normal while this is going on.

Now we are hearing about "hidden heat" deep in the oceans and changes in global wind patterns. It's flabbergasting watching all this. First of all, Dr. Bill Gray has pioneered the cycles of the ocean known as the meridional circulation for over three decades. I have linked to his paper several times. But when the tropical Pacific changes from warmer to cooler, it has a huge effect on the global wind, increasing the easterlies (trades). How this is somehow being offered as an excuse is beyond me. You "discover" something that has been known for years, and then you tell me it's because of global warming?

Here is an inconvenient fact: When water sinks (downwells), it warms. That so-called warming is a sign of a large-scale oceanic circulation, which has been occurring since way before man was here and will continue way after.

You know, all this would be funny (actually it is) if we already weren't seeing the effects of an energy policy based on non-facts. I wanted to get another tank of propane here, which normally cost $250. They wanted $700. When I balked, I found out that the propane shortage has reached the eastern U.S. I just had my highest electric bill ever, and that was from December. Lord knows what it's going to be in January. The economic toll this winter is taking, plus the lack of understanding how cold it can get, has left our nation in a very bad position. (Remember five years ago when we were told, "Children are not going to know what snow is"?) Just what do you think will happen next year when coal plants go offline? We will be left out in the cold — literally.

But one has to keep their sense of humor about all this. I am trying to be a happy warrior rather than a ranting worrier, but it's a battle. It's tough to smile when you realize that what people are pushing has nothing to do with the reality of cold, hard facts.

The Climate Chronicles

When dealing with things like this, you have to figure out what is going on. I wrote "I've Figured Out the AGW Strategy" on March 14, 2014.

I am starting to think the AGW/climate change/climate disruption/carbon pollution agenda is to simply make such outrageous comments that my side gets tired of having to deal with it. This hit me a couple of nights ago when I tried to watch five minutes of the "all-nighter clim-a-thon" being staged by a select group of senators. After a few minutes—and I will not say what I heard or who said it to set me off—I decided it was enough, so I watched something with more substance—"Dallas"—on TNT.

Oddly enough, Bobby Ewing stopped his nephew from fracking his part of Southfork by saying there was an endangered species on the property (the lean Prairie Chicken, I think it was). Even Dallas has an energy-based debate going on. Interestingly enough, Southfork, which in the show is a working cattle ranch, is also in the crosshairs of climate change since, after all, we all know cows cause increases in methane, which is one of those dreaded greenhouse gasses that, if left unchecked, will take Earth to Venetian temperatures (sarcasm).

And besides, eating meat is bad for you. I never have steak for dinner anymore—it's usually breakfast, when I can afford it. (Which is getting rarer. Then again, that is how I like my steaks: rare).

So when I was writing this week, I decided to have a little fun and pull out an outrageous example of how absurd this is getting. And I am not going to pick on the obvious—some of the leaders of our government pushing all this. Instead I want you to think about this headline from the UK Daily Mail:

"Will climate change bring back SMALLPOX? Siberian corpses could ooze contagious virus if graveyards thaw out, claim scientists."

This is fascinating for a couple of reasons. Just how long has Siberia been warm where all these corpses are thawing out? Well, over the last three years, it hasn't been warmer than normal except near the Arctic Circle, where it's so cold anyway that a one-degree rise isn't that big a deal. Besides, who is buried in the Arctic Ocean? (Maybe that is where Jimmy Hoffa is...)

And if they were thawing before, they certainly aren't this winter!

But this is not the real question. Again, think about the extremity of foolishness going on here. Climate alarmists are so foolish that they don't even stop to think: *Just how did they bury the people in the first place if the ground wasn't thawed out before?* It had to be warm enough to dig deep enough to bury these people — or did they have some type of drill none of us know about during the time of small pox? The whole article supports the notion of *natural* climate variations, because if it warms enough to thaw out these graveyards, it is merely warming to where it had to be before. You can't have it any other way.

But things like this are all too common in this argument. In all my years in weather and climate, which I use as a major piece in the foundation of my forecasts, I have never seen such nonsense and ignorance being spewed with no regard for the facts of what actually happened.

I am on everyone's mailing list, so I get sent countless articles on this. I also get sent the scholarly articles both for and against human influence on the climate. I know and understand the reason this is actually a debatable point and have pointed out several times that two major indicators — stratospheric *cooling* and the outgoing radiation measurements — support the idea of a slowly warming troposphere. Whether that is natural or man-made is impossible to know, for the earth has warmed before (it had to if you could bury people in Siberia) and has also cooled. How one could possibly know this is CO_2-driven given the history of CO_2 and temperatures that show no correlation, yet alone causation, is beyond me.

But I guess one needs to keep a sense of humor about all this. I believe many of the climate alarmists are whistling past the graveyard.

In the meantime, just in case they are right, I have canceled plans for a tour of Siberian graveyards. It's too cold anyway, given the actual weather.

Of course, one of the biggest odd means to an end is Earth Day, so I penned "Reflections on My 45th Earth Day" on April 23, 2014 (maybe it was 44th — you lose track of them).

I remember the first Earth Day in 1970. I was a freshman in high school and we had an Earth

Day Assembly. There was a play with Steppenwolf's "The Monster" blaring. They changed the words "America where are you now, don't you care about your sons and daughters?" to "America where are you now, don't you care about your *parks and waters*?" The young ladies from the drama club and the cheerleaders were dressed up as trees running around on stage. When I saw that, I wanted to become a tree-hugger, but alas, none of the trees was interested in hugging a 5' 2" 160 lb. freshman nerd.

Years later this has grown into a borderline religious holiday for some. I thought Earth Day Eve caroling would be a good idea until I was pelted with recycled cans and organic waste. And getting dressed up as a tree was not a good idea since every dog in the neighborhood was howling at my lousy singing and trying to use me as a place to relieve themselves.

Thank you, thank you very much. I'm here till the editors get tired of me. Try the Tofu Veal.

Given the myth and reality of what is going on with our planet and the evolution of this movement into some kind of socio/political/economic/religious/intolerant machine, I thought I would infuse some humor into the situation while reminding people that the planet is not in danger of becoming a wasteland by what we hear from these people. Other things, perhaps, but not global warming. There are far more pressing problems facing mankind today, chief among them the longstanding one of man's inhumanity to man, which seems to be the root cause of many ills. But when I look at some of the things that are being spouted as evidence of environmental disaster, it's easy to see how this issue is far from the end of existence it's portrayed as. The most visible high priest of Gaia is perhaps Nobel Prize winner Al Gore. Yet the doom and gloom since his Academy Award-winning movie "An Inconvenient Truth" came out has burned up like paper in fire. The movie was a major accomplishment, though whether you wish to admit it or not, it did for the global warming movement what "Triumph of the Will" did for Germany in the eyes of many in the 1930s. How ironic that we find people on my side of the climate issue being demonized with terms that bring up memories of arguably the greatest case of man's inhumanity to man by the very people who recently put out this movie. Yet the inconvenient truth in both cases is that the missive was proved false.

But let's try to keep a joyful warrior spirit. The use of facts is convenient for that.I will use some of the more choice worries from "An Inconvenient Truth."The melting of the ice caps? *Global* sea ice in 2014 was above normal.The Southern Hemisphere was breaking daily records and threatening to reach all-time high anomalies!

Earth does this back-and-forth act in all things. When some places are warm, others are cold. Because the oceans have been in their warm cycle and this affects most the northern ice cap, it has been well below normal. It does so by warming the land masses of the Northern Hemisphere, which have greater temperature ranges, and by doing so, influences the Arctic, a landlocked ocean. But during its heyday the Southern Hemisphere was *below* normal at the start of the satellite era. The testable theory is that once the Atlantic shifts to its cold cycle in the coming five to 10 years, the Arctic ice cap recovers. It's then that climate alarmists will try to shift attention to the shrinking southern ice cap. (At least it better be shrinking. Note it has started and the response on the northern side is record snow and ice recovery on Greenland.) That is the whole natural cyclical theory for ice caps — when one expands, the other shrinks. So it's a test. And yes, the Arctic ice cap has decreased overall since 1978, but you don't need to be a math major to understand that if *global* sea ice is *above* normal, it means that there is a greater compensating *increase*, counter to the missive we have heard for all these years.

Then there is the issue of hurricanes. Again, aren't we looking at this globally? Since the movie and the dire pronouncements, global tropical activity has sunk to record lows! The ACE index shows there has been no hurricane attack since the release of "An Inconvenient Truth." **[It took until 2017 to get another major hit. And back in 2014, the prediction below was made, and so 2017 was no surprise.]**

This should return toward normal and even go back above in the coming years. Why? The simple answer is because nature swings back and forth in cycles. You see, in spite of trying to make this more complex (certainly the details are more complex), the main missive remains the same: The earth by its design has no "perfect" climate but is in constant search for a balance it can never attain.

Then there are tornadoes. For the third year in a row, after the one major year of 2011, we see abnormally low tornado production. It should pick up next week, but as of this writing we are at record lows for the date.

Then there is the 10-year running global temperature since the Pacific began entering its colder mode. Since the year of "An Inconvenient Truth," the inconvenient truth is temperatures went down for a while, then spiked with the 2017 Super Niño and are falling again.

Regarding all those computer models that had so many in a tizzy about the impending doom of the planet, basing policy on computer modeling is a fool's errand since they busted so badly. People who forecast every day understand that no event is true until it actually happens.

We can go on and on here. The fact is this: Given the actual geological record of the earth's temperatures vs. CO_2, it's *cherry-picking to use the intervals of warming* in the past century to claim it's man causing it.

As far as the Hockey Stick goes, I count around 100 scholarly articles showing the earth has been warmer before. How is it we are to trust one without even being able to see the data behind it?

As I sit here writing this reflection of Earth Day, I will end with this. I believe the intent of the original Earth Day was good even if the cheerleaders dressed up as trees would not allow me to become a tree-hugger back in 1970 (they probably couldn't get their arms around me anyway). I think that being good stewards of this garden called Earth (I don't believe it's Gaia, but something that is given to us by God, and as such we should take care of it) is spot on. But to deal with reality, one must face reality, and there is demonstrable evidence that runs counter to the missive this has grown to now. And to show you what a good sport I am, I like the movie "An Inconvenient Truth." But I like a lot of movies where fantasy is involved.

It's up there with "The Wizard of Oz," a movie in the day where tornadoes were not caused by man as they try to make you believe now. But given my love of the weather, saying I like "An Inconvenient Truth" the way I like "The Wizard of Oz (I like "Star Wars" too!) is high praise indeed.

Hope you had a holly jolly Earth Day.

I wrote "Fossil-Fueled Fiction" Oct. 20, 2014, because I had some nice comments to make to show how this is so distorted.

It's been a heck of a week for hysteria in the so-called Climate War. Apparently, with all the other things going on in the world, the hysteria has to be whipped up so people actually pay attention to this trumped-up agenda.

Look at this wattsupwiththat.com headline, an inconvenient truth if hysteria is your agenda: "World Disasters Report for 2013—lowest number of catastrophes and deaths in 10 years."

That's right. All this screaming and yelling about how bad it is, and we find that we are at a decade low in terms of catastrophes and deaths. Wait a minute — isn't this extreme of non-extremes an extreme in itself? This lack of disasters is a disaster since it's a sign there have to be more disasters. After all, how can there be any less? Don't be surprised if that argument shows up. After all, only in the world of the AGW alarmists can a record-breaking amount of ice in the Southern Hemisphere and the third-highest snow total for *September* in the Northern Hemisphere be a sign it's getting warmer. Apparently, humans are also changing the freeze process. Perhaps water now freezes at higher temperatures, and that's the reason increases in snow and ice are a sign it's warmer? (Sarcasm.)

And then we get this from the founder of the "#ClimateSilence" and "#DontFundEvil" campaigns, Brad Johnson, who calls himself "Climatebrad": "Dangerous, Fossil-Fueled Hurricane Gonzalo Barrels Toward Bermuda."

Is that so? Gonzalo went through the area we said in April would be primed for the strongest storms. We predicted one or two major hurricanes.

Guess what? Gonzalo was the second major hurricane of the year.

Apparently, this "fossil fuel" didn't work too well. It *weakened* from Category 4 to Category 2 as it reached Bermuda—quite a bit different from the non-fossil-fueled Hazel in October 1954, *which hit the North Carolina coast as a Category 4* and caused hurricane force winds all the way into Canada.

Moreover, the hurricane season this year has gone almost exactly as our non-fossil fuel-based forecast said it would when we put it out in April! The storms went where we had the two prime regions for storms. All of the hurricanes had hit maximum intensity in our two main areas. We had little activity in the deep tropics. If there is any criticism, it's that the area of highest threat should have been centered east about 300 miles. But without any of this fossil fuel nonsense, this forecast from April targeted the season we have had. We are within one now of the total number and all the *hurricanes* (Dolly was tropical storm) have gone through our areas. If we get a late-season development later this week, it reaches the low end of the total number. The forecast had one or two major hurricanes, and we've had two form. As of Sunday p.m., we are near 75 percent of total ACE.

I could not possibly have had the life I've had or do what I do (this applies to all of us) without fossil fuels. Which is one of the things that bugs me so much about the whole climate alarmist movement. Here we are, all nice and warm with our marvelous technology, and the same people and industries that help so much to make it that way are now demonized and destroyed. I believe the word to describe someone that does that is "ingrate."

Just a reminder folks: The true "hockey sticks" of fossil fuels are not with temperatures but with worldwide GDP and life expectancy.

Though showing you only a couple of examples, it's obvious the strategy of the AGW alarmists is to simply blame everything on fossil fuels. Even in the year of a lack of extreme events, they are trumpeting all weather events as evidence they are right. In a way, they are playing an "everybody gets a trophy" card ("no matter what we say, we are right, give me my trophy") and a victim card ("the evil fossil fuel people are destroying us; never mind we are far better off than we would have been if what we are pushing was enacted from the get-go").

I guess then it's okay for me to call what they do what it is: FOSSIL-FUELED PROPAGANDA.

Why not? They claim everything else is.

It gets so bad that one might think, "Is there a CO_2 fairy waving its magic wand?" While CO_2 has little to do with actual weather and climate, as shown by past events both recent and in the geological time scale, apparently it can affect people who believe it does.

Side note: Look at the 2017 hurricane season. The CO_2 fairy was causing a massive upswing in the Atlantic basin but fell asleep in the Pacific until late in the season, when two or three typhoons went off. Never mind the patterns we have researched for decades showing that this kind of thing was going to happen — it must be CO_2. Capture the CO_2 fairy and you rule the planet.

"The Wonder of It All" was written on June 24, 2015.

I was thinking about the events in the wonderful world of climate over the past week, which as usual is more about what someone said rather than anything that happened. I am trying to make sense of it. For example, why is the hockey stick-like connection between life expectancy, population and per capita income not credited to increased CO_2 brought on by the fossil fuel era? It is not by magic that humans have it better off now than we did before fossil fuels came onto the scene.

Yet somehow temperatures are showing that, whatever effect CO_2 has, it has to be rendered so small that one would be hard-pressed to find any significant linkage.

That's not to say CO_2, which comprises 4 percent of the greenhouse gasses and .04 percent of the atmosphere, has nothing to do with the wonderful 33°C of warming that greenhouse gases supply. It is to say it is so small against all the major forces that affect climate that it's difficult, if not impossible, to see the effect in the factual history of CO_2 vs. temperature. So why is it suddenly now doing all it's purported to do when it has been much higher in colder times and lower in warmer times?

Not to wade in too much, but why is 2 percent of the pope's encyclical attracting either the backing or attacking of it while the other 98 percent is not even referenced? I wonder what a lot of the backers would think of some of the other issues in the encyclical, since most of what is in there contradicts a lot of their world view. That is the wonder of it all. A world obsessed with a thermal hockey stick ignores other hockey sticks that are more indicative of the human condition. But here's the biggest wonder of it all for me: How did something I have loved for over 55 years get twisted into what it is today?

Which leaves me with one more question:

Will wonders never cease?

Here is something I wrote when I worked at Accuweather and republished at *The Patriot Post* with Accuweather's permission. I did so because it is my dad's favorite of all I have written. But I think in having a little fun we can see that perhaps things are not so different as far as people's behavior. Only the magnitude has increased.

On Aug. 8, 2016, I wrote "To Dad, From Joe."

> My father reads everything I write. Most of my weather blogs are written very quickly as I need to get information out in a timely fashion. So my mom will print them out and my dad will correct them with a red pen. I guess there are boxes full of corrected blog postings as a result of this. *The Patriot Post* sharpens up my commentaries, so I don't know how much red pen it uses on them.
>
> But there was one piece my dad always loved — a takeoff I did on "Julius Caesar" — and my mom mailed it to me the other day. Perhaps you might enjoy it too. So for my father, and in the spirit of *Mad Magazine* I used to read as a kid, here it is.
>
> **If Caesar Were a Climatologist: Marc Antony's New Speech**
> (With apologies to William Shakespeare)
>
> Since some people believe the earth is as warm now (or warmer) than 2,000 years ago, this may be worth the read for those with a sense of humor or maybe, just maybe, for those who

still have an open mind on the natural variability of the earth's climate.

There are people who believe that parallel universes exist. One may have existed around the time of Rome when the climate was warmer than today. The Romans could have been concerned about Caesar's stubbornness in dealing with the issue of global warming and took matters into their own hands. So what follows is a parody of Marc Antony's speech in Act 3, Scene 2 of "Julius Caesar." In this world, Caesar was a climatologist and was apparently at odds with the climate models and their backers. The text deletes the responses of the crowd and deals with Antony's words only.

From "Julius Caesar":

> Friends, Romans, citizens of the world, lend me your laptops and iPhones
> I come to bury the notion of cyclical warming, not to praise it.
> The evil of global warming lives well after CO_2 is spewed into the air
> The good of adaption to cycles is never brought up.
> So let it be with the ideas of Caesar. The noble Climate Modelers
> Hath told you Rome is causing global warming
> If it is so, 'tis a grievous fault,
> And grievously are those believing different paying for it.

Here, without the models and their creators
For the Climate Modelers are honorable entities
So are all who follow them, all honorable men
Come I to speak at the funeral of Caesar.
He was my friend, a great climatologist and meteorologist
But the Modelers say he was stubborn
And the Modelers are honorable and always right.
The system Caesar had brought much to Rome
Whose people prospered like no other.
Was this in Caesar causing the warming?
When there was no air conditioning, we created it.
It saved lives. Did this cause too much warming?
But the Modelers say Caesar caused global warming
And the Modelers are all honorable and always right.
You did see the evidence of warmer times
It happened once, twice, thrice
And thrice it turned cooler again. Is this global warming?
Yet the Modelers say we cause the warming
And the Modelers are honorable and always right.
I speak not to disprove what the models say

But I am here to speak what I do
know
We all loved our way of life
once, not without cause
What causes you then not to
mourn for its passage.
O balance of ideas, thou art fled
to brutish beasts,
And men have lost their reason.
Bear with me.
My ideas are in the coffin with
Caesar,
And I must pause till they come
back to me.
Yesterday the words of the past
experience might
Have stood against the world.
Now there they lie
And no one does them
reverence.
Oh readers, if I were disposed to
stir
Your hearts and minds to
reading the facts
I should the models, and the
Modelers wrong
But you know, they are
honorable and always right.
I will not do them wrong, I
rather choose
To wrong the past, to wrong
myself and you
Than I would wrong the
almighty honorable Climate
Modelers.
But here is a parchment with the
seal of Caesar

I found it in his closet, 'tis his
doctorate on Cyclical Climate
Change
Which I do not mean to read.
And they would go and kiss
dead Caesar's wounds
And dip their napkins in the oil
of his SUV
Yes beg a hair of him for their
memories
Which can be beautiful and yet
What's too painful to remember,
we simply must forget.

*NOTE: Apparently Barbara
Streisand assisted Shakespeare
in this parallel Rome. How else
could a line from "The Way We
Were" show up?*

Have patience gentle friends, I
must not read it
It is not meant you know how
Caesar studied this.
You are not wood, nor stones,
but men
And being men, reading the
thesis of Caesar
It will inflame you which could
lead to more human-induced
global warming.
'Tis good you not know what
his study said
For if you should then what will
come of it?
Will you be patient? Will you
stay awhile?
I have o'ershot myself to tell
you of it

I fear I wrong the honorable Climate Modelers
Whose result stab Caesars Natural Climate change thesis.
You will compel me then, to read his thesis
Then make a ring around the classroom
And let me show you what is in his thesis.
Shall I open this and then will you give me leave?
If you have tears, prepare to shed them now
You all do remember Caesar's work on climate.
The first time he discovered cycles
'Twas on a hot summer evening, in his tent.
The southwest wind bringing in an air mass from Africa
Across the Mediterranean around the back side of a heat pump of high pressure.
Look, in this palace was one Modeler ripped out that passage
See what another did in trying to adjust his data.
Through this page, one of his students turned on him at a conference,
And as he twisted the date to fit the model
The thesis of Caesar was being torn to shreds.
And defending his ideas with evidence of past cycles

If the Modelers unkindly disputed or no
For the modelers were something he welcomed.
Judge how open minded he was about new information
And that was the most unkind cut of all.
For when the modelers he supported stabbed him
Ingratitude, more strong than the traitors arm
Quite vanquished him, then burst his mighty thesis.
And no matter how he brought up the facts
Even at the base of Pompey's Statue
Which all the while knew he was right, his thesis was ripped.
And oh what a rip that was my countrymen.
There I and you and all the things we relied upon fell down
Whilst new age "latest science" flourished over the hard work of decades.
On now you and see and I perceive, you fell
The dint of pity, these are gracious drops.
Kind souls, what, weep you when you but behold
The ideas of years and years in Caesars thesis
Here is, marred as you see, with model mayhem.

(Now knowing the facts, the citizenry erupts, but Antony speaks anew.)
Good friends, sweet friends, let me not stir you up
To a flood of mutiny.
The Models and Modelers are honorable, and always right
What incorrect data they have, alas I know not
That they say these things and are always right, and honorable
And will no doubt, with time verify
I come not friends to steal your heart away
I am no Modeler, and not high tech
But you know all me, a plain, blunt operational meteorologist
That loves the weather and that they know full well
That gave me a chance to speak of the cyclical warming of Caesar
For I have neither the math, nor the models, nor the funding, action, nor utterance, nor the power of Hollywood
To make movies on global warming I only speak right
I tell you of which I do know
Show you the ideas Caesar had, poor poor dumb mouths
That I was a Modeler, and had a model
That showed the cycles to be as they are, facts

The stones of Rome would rise
up and mutiny
Yet hear me country men, let
me speak
Well friends you go to do you
know not what
Wherein hath Caesar thus
deserved your love
Alas you know not, so I must
tell you
You have forgot the thesis I told
you of
Here is the thesis on climate
change
That it has been warm before, it
will be warm again and it will
cool
From time to time. Moreover it
will warm even further
Before it cools again, so that is
what has happened
Cycles in the oceans, solar
cycles, volcanic activity
Are just a few of the many
items that can determine
Our climate. We are in a warm
cycle now
Why do we have palm trees all
over the place? How have
Our armies conquered to shores
once covered in ice?
Just how did men in togas get
over the Alps?
Have you ever seen a Roman
fight in an arctic parka?
We don't have too
Cause it's warm, as war as it
was before, but years from now
should

Rome not wake up and use a bit
of common sense
The empire will fall and the
snows will return, all the way to
Rome
That is the thesis of Caesar
Here was a climatologist. When
shall come another?

Unfortunately, people who have odd means to achieve their ends are not going to simply play fair or even smile. As I said, an open hand is often met with a closed fist. And while I have pointed out some things in this chapter and tried to have a bit of fun, the fact is when you see things like this, you realize that what you love and cherish is being attacked.

"A Chilling Shot Beyond Climate," March 11, 2016:

A headline from The Blaze I ran across this week: "AG Lynch Testifies: Justice Dept. Has 'Discussed' Civil Legal Action Against Climate Change Deniers."

At the age of 60, I simply cannot believe that this can even be considered in the nation I grew up in. If the climate issue is such a done deal, why are we spending so much money to tell everyone it is a done deal? And why would anyone in authority try to silence people who disagree with them if we are so absurdly wrong? At the very least, this was designed to intimidate, and it's hitting its mark.

To people who don't see this issue the way I do, or see things different politically, is this what you really want?

There is a line in the Bruce Springsteen song "Drive All Night" where he is questioning the boundary of how tough he really is: "I wish God would send me a word, send me something I'm afraid to lose." I always listened to that with the understanding that it's only by overcoming fear that you truly know courage. All my life I thought that applied to wrestling, training, endless hours of work, or trying to support my family. Never have I thought it would be something like this, where my fear would be of the nation and system that I fell in love with. But I guess the good Lord will send such things to reach into the depths of a man's soul to bring out the best in him.

Things have changed. It used to be: What would you do to defend your country? But the fact the silencing of climate skeptics is even being considered means you have to think: What would you do to defend yourself from your country? And that is scary.

And more than that, it's not what this nation has ever been about.

In the end, the above line sums it up. This situation is about things bigger than the movement of a few tenths of a degree. It's about absolute truth versus relative truth. It's about belief in things bigger than you versus a belief that you are bigger than things around you. It transcends science to ask the bigger questions about what you believe as far as man's role. We are ending with something that in many people's hearts and minds is the biggest question. The piece "There's Nothing New Under the Sun," written around Christmas 2013, acknowledges a higher power and offers lessons that have been eternal which are now being challenged.

I thought I would bring back a golden oldie to counter the revisionist history that is being pushed in the climate debate today. The same outlet that is now actively touting man-made global warming (which one isn't?) was not singing the same tune in the 1970s, and the *Time* magazine cover from April 9, 1977, proves it. It offers advice on "How To Survive The Coming Ice Age."

In the spirit of rallying individuals to counter a common threat, the sub-headline reads: "51 Things *You* Can Do to Make a Difference." (Emphasis added.) This is not unlike the activism being advocated during World War II (victory gardens, collecting cans, etc.). In a way, it's a noble appeal to the individual, which I have no problem with, except it makes about as much sense as all the noble appeals today to counter global warming.

There is always that appeal to the individual. Isn't that interesting? But there is a problem when the collective decides to *force* the individual to do whatever the collective wants.

Another fascinating link: the idea of uniting mankind behind a common cause (climate) in the name of building peace — something that was first put forth by Woodrow Wilson who, in some circles, is the father of the progressive movement here in the United States. It is very interesting that this appeal was made at a time when another energy crisis was developing, which prompted the sitting president, Jimmy Carter, to propose conservation legislation, including a windfall profits tax on oil. That, of course, collapsed when the price of oil tanked several years later. Apparently, there are cycles in the economy and climate (who would have thought?)—neither of which the president at the time seemed to have a handle on. (The reader can fill in their own ideas on the situation today.) But part of the reason for the "crisis" is similar to today: dependence on an outside source for our energy lifeline and the climate. In that case it was a very real demand because of all the cold — the same cold that prompted people to claim an ice age was imminent. That they deny it today speaks volumes, when, like past climate events, all one needs to do is look to see that there is nothing new under the sun. This time though, it's climate hysteria that is causing the impact on the market, as it limits what the nation can actually produce. One of the other ways is to simply manipulate data; to try to get rid of the evidence—a scientific version of book burning.

So here we are—another climate crisis, this time replete with hysteria over warming that is perfectly within the realm of what nature is supposed to do, much like the hysteria about cold was. But with it comes more demands to do things to curb the climate, etc. It's noble to try and unite mankind in a peaceful cause by appealing to the individual. I would like to think that is the true motive here: good old-fashioned American ingenuity and drive to conquer a global problem for the betterment of mankind. But what happens when the problem 40 years ago was opposite of what it is now, yet under the same overall heading? And what if the attempt at conquering the problem was based on a motive that had nothing to do with actual facts? The reader is left to decide. But the fact that cold and warmth are being blamed on the same thing is simply a case of someone covering all their bases and claiming all their answers are correct, even though that can't possibly be the case.

In the end, what has happened before will happen again. There is nothing new under the sun. That goes for trying to manipulate people as well as the weather and climate. All one needs to do is look at what has actually happened and the whole agenda is clear to see.

It only seems fitting to conclude with Ecclesiastes 1:9: "What has been will be again, what has been done will be done again; there is nothing new under the sun."

The absolute truth of nature and nature's God will always win out over the relative truths embodied by the idea that the means (odd ones) justify the end, if truth is the metric. If not, then you reap what you sow.

Chapter 7

Why Nye?

I am in a movie with Bill Nye where I am portrayed as a counterweight who apparently wants to thwart his life's mission, which is to save the planet. Besides the obvious differences of opinion we have on man-made global warming, our lives are also opposite as far as mission is concerned. Bill is out to change and save the planet. He says so. I am not out to do that. If I am right and it helps, fine. But the reason I am the way I am is because I believe we are just vapors in the wind and represent a very tiny part of a majestic Creation. We cannot change nature; nature controls us. That we think we can be bigger than nature is arrogant to me. I have come to believe that whatever talent I possess was given to me by God and that my affection for weather eventually made me grateful for Him. Whatever I do, I reach toward Him. Obviously, saving the world is beyond my pay grade.

It's not unlike what we were taught in wrestling, which is almost a spiritual thing. Constant preparation, focus on process and a foundational base all come together to produce the result. When praying "Thy will be done," you have to mean it. It's not "My will be done." This is not an excuse to say, "Oh well, too bad I was wrong" or "Oh well, too bad the planet is burning up and I am stopping Bill Nye from saving it." If you focus on outside goals, and if the objective is not the pursuit of truth, then your judgement will be clouded.

All that is to say: You must travel with no baggage. Luke 9:3 says: "Take nothing for the journey — no staff, no bag, no bread, no money, no extra shirt."

Baggage in this case is the idea that your answer *has* to be the right one.

Stay focused on the pursuit of the correct answer. When Peter looked around at the waves, down he went. If I have to worry about what this person says or what that person says or how bad I get slammed in the media if I am wrong, be it anything, I reduce my chance at finding the right answer. If I get the right answer, then it helps people who heeded it. The worry about what people say or do is like staring at the waves.

But Bill has a tremendous weight on his shoulders, and it's certainly something that could blind him. I have explained this already in this book by debunking the whole missive of the climate movement. But the idea that somehow my path has become linked to some extent with Bill means I should explain some things.

This all got started with a 2010 debate on "The O'Reilly Factor" that I never really wanted to do. In fact, I am tired of three-minute sound bite debates. Bill, in a recent PBS trailer, said I originated the challenge. I did no such thing. O'Reilly's team called me up and asked me to do it. I was reluctant because of my kids. They said, "Dad, this is not like a regular debate. To kids, Bill Nye is the Easter Bunny and Santa Clause. You can't go after him." In the end, I did the debate, but if I had known Bill was not the Easter Bunny and Santa Clause I would have not simply been smiling and having a good time. Bill loves to reference Venus with its CO_2 concentration, and he did so here. We've got a long way to go before we are on par with Venus, which has substantially more carbon dioxide. But it's a typical debate tactic — bring something up that doesn't specifically deal with the matter at hand. The point is that somehow this debate ensnarled us.

I ended the debate with a challenge to which he never responded: What would the temperature do in the next five years? If he had said it would go up, he would have lost.

So in an article I penned years later I wanted to use the coming El Niño to demonstrate the direct correlation it has on global temperatures, independent of CO_2, which guys like Bill blame for everything. To do so, I wrote an article revisiting that debate.

Now, I have done several debates and, as I said, I am not doing them on TV anymore. I truly enjoyed the one with Stephen Colbert, and he was great to me. In fact, he came up to me before the show because he saw me working and said, "Can't you just relax for a while and have a good time?" He knew I was a workaholic. But a warmup guy and I were all laughed out by the time I was up. I met the "Reverend" Al Sharpton, who had his own room. His hair, as Zevon said in "Werewolves of London," was perfect. That was fun.

I had a debate in 2015 with former Rep. Dennis Kucinich on Sean Hannity's show. I trapped him on a question on the January 1978 snowstorm in Cleveland, a much stronger storm than the one we were debating in New England. He claimed it was the result of lake effect, which it was not, and asserted that local Cleveland TV legend Dick Goddard told him it was. I called Dick up the next day, and he said he was sure Kucinich misheard him. Kucinich is a nice guy, by the way, and I confess I liked some of the things he would say in other venues. But I didn't like that particular notion. And then it hit me — if you have an audience and they don't know the facts, and someone claims, "Yeah, it was caused by lake effect snow," the audience would simply say it's a tie. And a tie on a debate involving weather while the real agenda marches on does me no good.

I had another debate on Hannity with a guy who runs a green-energy-based hedge fund. He said he had no dog in the fight and changed the debate to ocean acidification. So you run a green energy hedge fund, but you have no dog in the fight?

In fact, in the movie I am in with Bill Nye, what I wanted to do was sit in the backyard with a glass of wine, like the

Italians loved to do in the olden days, and discuss the issue, because, truth be told, that is where this debate should be held.

Think about this: Bill has five million Twitter followers. That is a lot of social media testosterone. I have less than 50,000. Why would he want to "punch down"? It baffles me.

In any case, I wrote "Some Questions for Bill Nye Six Years After Our 'O'Reilly Factor' Debate" in November 2015.

This was written not to entrap Bill but to get it out there that temperatures had not gone anywhere since that debate, and, *absent any use of CO2 as a tool*, I was forecasting the major El Niño and the ensuing temperature spike. I do this quite often on social media — blazon an event I know is coming, make the forecast *and trot out the projected reaction* of people who had no idea what was coming.

> This Salon article caught my eye because during the last El Niño I was on "The O'Reilly Factor" debating Bill Nye:
>
> • "Bill Nye demolishes climate deniers: 'The single most important thing we can do now is talk about climate change.'"
>
> At the end of our debate, I challenged Bill to a grand experiment, something a man of science like him should love. I put forth an idea on where global temperatures would go by 2030. I opined it would return to the level, as measured by the National Center for Environmental Prediction (NCEP), it was in the late 1970s, when we began recording real-time temperatures.
>
> You may have noticed over the years Bill's increasingly shrill tone on this matter. I haven't

followed his career all that much, though I knew before I debated him that he and I did not share the same opinion on global warming. My kids, then 14 and 11, pointedly told me I was in essence debating a man as beloved as Santa Clause and that I should be "nice."

I want you to read the whole article above. Here is a man who *never responded to my challenge* now saying this:

"Part of the solution to this problem or this set of problems associated with climate change is getting the deniers out of our discourse. You know, we can't have these people — they're absolutely toxic."

Is this the way a man of science speaks? Label people who disagree with him as toxic? Then again, in spite of obtaining a degree in engineering from Cornell, the fact is he is not a man of science. *He is an actor.* That is his profession. We shared the same math and physics classes as a foundation for the core of our majors, but the big difference is that I have worked almost *40 years* in meteorology and made history a foundation of my forecasting methodology, while he has become an actor. A person with science as his driving motive does not refer to people who disagree with him — such as 30,000 degreed scientists, over 9,000 of whom have PhDs — as "toxic." Only a man with other motives would seek to isolate, demonize and destroy those who disagree with him — something you see out of the Saul Alinsky book *Rules for Radicals*, which I have read. Please read Saul Alinsky's rules and ask yourself: Is that not what is being used and personified by Bill Nye?

This year another El Niño is spiking global temperatures, so I will challenge Bill to confront three simple facts.

1) Please explain the lack of linkage between CO_2 levels and temperatures in the established geological record of the earth.

2) Since the debate, explain why temperatures have not risen. **[Again, the El Niño was coming, so I wanted to get the reasons out beforehand.]**

3) As stated, we are in another period of increased warming. Since Bill won't take part in the grand experiment I suggested, perhaps he can tell us where global temperatures, as measured by NCEP data, will be one, three or five years from now.

To help him out, I have it warmer next year at this time than now, colder than 2016 in 2017 and temperatures in 2018 and/or 2019 at or below their lowest point in 2012. **[2017 is running cooler than 2016 and La Niña has developed.]**

If Bill concedes my point and actually makes a forecast, rather than call people like me toxic for challenging him, then he will be saying that the 14-year period ending in 2019 would have no temperature rise and even a fall! That would mark over 20 years with no rise. Given the increase of CO_2 of 1.8 ppm a year, that would mean we increased carbon dioxide levels close to 10 percent since the late '90s, but with *no increase in temperatures.*

If he says the ensuing years will be warmer, then finally we'll have him on record — even though temperatures fell without him participating in the original challenge in 2010—and we can see who is right and wrong, as I have to do every day in my job. Apparently, no such standard exists for actors masquerading as scientists. If he challenges the scale, then he challenges NCEP. By doing so, he implies that its system for measuring temperatures—something essential for model initialization—is wrong. If so, then Congress, instead of investigating climate skeptics under RICO statutes, should investigate NCEP for all the money it has spent developing models that, according to Bill, are wrong. (Note the sarcasm. NCEP temperatures, the gold standard of real-time temperatures in my opinion, are just fine.)

So to reiterate these three simple points:

1) Explain why there's no linkage in the entire known CO_2-temperature history of the planet.

2) Explain the lack of warming in real-time temperature data and why I have been right so far.

3) *Make your forecast.* You claim to be a leader yet refuse to take a stand. Instead you sit in the stands and never allow what you are saying to be verified. What kind of science is that?

There's no need for another debate, though, for as Bill so eloquently says, "Part of the solution to this problem or this set of problems associated with climate change is getting the deniers out of our discourse."

Real nice, eh? Let's silence free speech and thought while we're at it!

I am not looking for another debate, since he has not answered the original challenge I set forth. I want his forecast! Put up or shut up.

I don't think any of this is toxic except to someone who refuses to confront simple realities and instead makes statements much more in line with an agenda-driven zealot than a man of science who's in pursuit of the correct answer no matter where it leads him.

After the El Niño drove the global temperature up, Bill did respond — in the spring of 2016. He waited until the warming I predicted occurred because his audience could not have known I forecasted the warming. I responded back with "A Real Bet for the Tough Guy in a Bow Tie" (ever notice how a lot of these guys love being superheroes?), written on May 3, 2016.

BREAKING NEWS: Bill Nye issued a bet more than six years after my initial challenge to him in 2010 (which he would have lost) and four months into 2016 after reviewing the impact of El Niño on global temperatures. News flash, Bill: Midway through last year I said 2016 global temperatures would rise thanks to El Niño. I can forecast this because I don't believe CO_2 is a major player in determining global temperatures. I believe the sun, ocean cycles and stochastic events play a much more significant role.

Just so Bill and the rest of his brainwashed audience understand, I fully support our nation's transition to clean and sustainable energy while using all sources of energy at our disposal now until a feasible economic transition can be accomplished. If you were really serious about it, we would be using more nuclear energy anyway.

However, I also believe the policies that Bill and the rest of the global warming political activists are pushing are detrimental to our economy and, in turn, our national security. How much money have we shipped to the Middle East because we did not use our own domestic fossil fuel resources? If not for the recent energy boom in the U.S., foreign countries would be making billions of dollars more at the expense of the U.S. consumer. I would argue that our failure to move more quickly and utilize our domestic fossil fuel resources has had catastrophic effects on our economy and national security.

Unlike Bill, I am a rational man, and I understand that while we must transition to clean energy, we must do so in a way that is smart and economically viable.

Furthermore, we all know that Bill is not a forecaster. And since I am, I have a bet for "The Science Guy." I believe 2017 will be cooler than 2016. The bet is this: For 2017, every increment of .05°C (plus or minus compared to 2016) will be worth \$10,000. If 2017 is 0.1°C warmer than 2016, I will pay you \$20,000. If 2017 is 0.1°C cooler, you owe me \$20,000.

We do it with Dr. Roy Spencer's satellite measurements.

The satellite data cannot be manipulated as we have seen in a culture among mainstream scientists. (Remember "Climategate"?)

Furthermore, since you say global warming is proven science, how about we take all the money allocated for AGW research and use it to improve veterans' health benefits. We wouldn't be allocating all that research money to study whether the earth is flat, would we? Or we could stop that AGW gravy train and use the money allocated to professors around the world for improving fusion output. Make sense?

One more thing. I challenge Bill to lead by example and for one year use no fossil fuels, including products derived from fossil fuels. He can be like the DirecTV commercial in which a settler is settling in a world void of fossil fuels.

See you Dec. 31, 2017, Bill. One of us will pay up.

2017 is running .07°C cooler than 2016.

All this lead to me being in a movie with Bill. Every superhero needs a villain, so I guess I am the villain.

The movie, "Bill Nye: Science Guy," includes this tease:

"A famous television personality struggles to restore science to its rightful place in a world hostile to evidence and reason."

I guess I am one of the people who are out to stop Bill from restoring science to its rightful place. For that I guess he wants guys like me imprisoned.

I didn't want to do the movie, but the oddest thing changed my mind. I am friends with two monster ex-football players. (Look in the dictionary next to the word "huge" and their picture is in there.) They are younger than me. I was at a charity event with them (Champions for Children in southwest Florida) and they scolded me for giving Bill a hard time. It hit me that my kids' generation all the way up to the generation behind me were influenced by Bill Nye. It also hit me that if two guys that big were into Bill Nye, then Bill Nye has to be huge. So that got me thinking. But it was the producers of the film who won me over. They just would not stop. I saw their previous work (I have to confess, they are good guys). They spent three days with me here, filmed me working, training, hanging out with my family, then went to Rhode Island to visit my mom and dad and all my relatives. They're down-to-earth guys. Then they came here again with Bill to film the wine summit on my back deck.

My goal: to show people what I am explaining in this book. I am a blue-collar guy who loves what he does, gives thanks to God above for it, and is simply pursuing the right answer. It's up to the producers to edit the film so it comes out that way. If it comes out that way, great. If not, whatever. It's not like I am that important. As a matter of fact, Bill has a lot more at risk if he doesn't look good.

I will say this. I saw a trailer for the movie on PBS. I think Bill overall is a nice guy. That may make some people mad, but I would not mind at all hanging out with him with no one around. But saying I am a climate denier is exactly what I am talking about with folks like him. An open hand is greeted with a closed fist. Just how is someone who has had success using climate and analogs for 40 years a "climate denier"? Just when has Bill ever made a forecast using climate and analogs that had to verify? If anything, he is a natural climate change denier considering that up until the past 40 years, CO_2 has never been the climate control knob. How can you be sure it is now?

I wrote "Fact-Checking Bill Nye" on March 1, 2017. In the piece there are three questions that must be answered.

> Let's fact-check Bill Nye regarding comments he made about something near and dear to my heart—wine—during the Feb. 26 edition of "Tucker Carlson Tonight."
>
> Here's his response to what the earth would look like today if man was not the culprit for all the warming:
>
> "It would have looked like it did in 1750. Britain would not be very well-suited to growing grapes as it is today. French winemakers would not be buying land to the north as they are now. People who plan to run ski resorts would still be able to do it in Europe."
>
> First of all, grapes are being grown in modern day Britain, but they were also being grown back when the Romans were in charge. There were 50 to 100 vineyards in Britain between 1000 and 1300 AD *and even one in Scotland!* And Britons were still giving it a go as late as the 1700s before the Little Ice Age. The reason is two-fold. 1) You could grow grapes and make wine because it was warm enough to (we were in a climate optimum both times). 2) The occupiers were of Roman and then French descent (the Normans showed up in 1066), and both are known for their love of wine. I can certainly speak for the former. Obviously, the northern tastes are more toward

ales, and this combined with the cooling lead to the necessary change from the fruit of the vine to the harvest of the field.

What about the ski industry?

There are plenty of resorts where people usually go to ski—the Alps. And locations north of the UK still have theirs.

I have a picture of two former pro football players, Chris and Keith Conlin, "choking" me because they wanted me to lay off Bill Nye. No, I am not going soft, because it's the totality of the journey that has to be looked at. But diminishing whatever authority you may have had before (if these two guys watched Bill Nye, then he has received a lot of exposure) with this kind of thing seems to be out of line to me. One is always taught to "finish strong." The argument here is not that the climate does not change — it always has and always will — but it involves, in my opinion, three main points that sum up my position:

1) How much is man responsible for variances that were previously exclusively natural?

In my opinion, most of the warmth today is likely natural given the tiny amounts of CO_2 relative to the entire system, of which the oceans have 1000 times the heat capacity of air and are the great thermostat of the planet, taking centuries of action and reaction to reach where they are today.

2) Is this worth the draconian reactions that will handcuff the greatest experiment in freedom and prosperity in history, the United States of America?

3) This question may arise: Would not the cost of adaption rather than trying to preclude an ordinary recurrence be a sounder fiscal response?

Now does that seem delusional or worthy of jail time?

I suggest Bill finish strong with his argument, but not by what is a soft tyranny of suggesting jail time or demeaning others, some with far more education and experience than him.

A big hint to all this lies in the cyclical nature of the oceans. The natural shift in the Pacific in 1978, called the Great Pacific Climatic Shift, from cold to warm started elevating global temperatures, and that was followed by shifts in the IOD (Indian Ocean Dipole). Most importantly for Europe and the Arctic, the Atlantic shifted to its warm phase. It does not take a science guy, just common sense, to understand the earth's major warm ocean bodies, with 1,000 times the heat capacity of air, would have a warming effect on the planet. (Inconvenient revelation: Warming oceans also release CO_2.) But why quibble with such trivialities as the ocean when you can just label people like the late Dr. William Gray, who had 50 years of experience in the field and a PhD—something Bill does not have—as

Former pro football players, Chris and Keith Conlin, "choking" me because they wanted me to lay off Bill Nye.

delusional?

Chris and Keith Conlin

Here's what it comes down to. By saying CO_2 is now the climate control knob, i.e. 100 percent responsible, you are *denying* (there's that word) the entire history of the CO_2-temperature relationship in the known history of the planet.

Also being denied is that warmer times, including when grapes were grown in Britain, were called climate optimums because life back then flourished.

And the increase even has a **diminishing** return.

I had the chance to talk with Bill Nye for about three hours in my home. He is not an evil demon; he is a man who believes he is right and wants to change the world for the better. That is what he believes. The problems for me start when he resorts to some of the things he has recently advocated — jailing people or, in the case above, calling people delusional. And that's a shame for two reasons. 1) It's the first step toward the type of despotic view that people who think they know better try to force on others. That is not freedom, and that is not science. 2) He is someone who really brought science to the forefront.

It's interesting. I recently read that Bill thinks he is a failure because he has not shocked us all into action. But Bill will get his way — because guys like me and the older generation that was taught to question is going to die-off. The problem is it will take a while longer, so Bill may not get to see his dream come true. Forty years from now, given low solar output, a shift in the oceans and a few volcanoes going off, it won't matter whether or not I was right, because the agenda Bill is pushing is similar to how the government views banks — too big to fail. There's too much wrapped up in it. I marvel at the fact that somehow I got hooked up in a situation involving someone who really did carve a niche for himself and is a household name to many.

So why Nye? I really can't say. But I would not like to bear his burden. Given what drives him, there can be no turning back. He is motivated to try to help, but the road to hell is sometimes paved with good intentions. You have to be able to say, "Wait a minute, maybe I am wrong." What you own owns you, and I guess I realize there is no answer I own. Bill is certainly striving to reach further, and he believes in what he is doing, That's fine with me, even admirable. I just don't force other people to adhere to my ideas. That is not admirable. To quote H.L. Mencken again, "The urge to save humanity is almost always only a false face for the urge to rule it."

Chapter 8

Top Five Blogs

Everyone loves lists, so here are my top five articles as determined by Facebook traffic that have not been included elsewhere in this book.

"A Gentle Reminder to People on My Side of the Climate Issue" was written on Dec. 13, 2016. The purpose of this piece was to ensure that people on my side of the issue remained focused on facts. There is a great temptation to fight the way the other side does — for instance, by using weather events as a sign you are right, or failing to keep things in perspective. I often use an analogy of a football team that is behind by five touchdowns and then scores a touchdown. It does not mean you are winning the game. For starters, the earth is warm. No one argues with that. But touting a temperature drop or an increase in Arctic ice relative to years when it's abysmally low as a sign it's not warm to me is tantamount to how the other side attacks this issue. I guess if my main mission was to win the real battle here (which is not against climate and science but another agenda), I would likely fight on those grounds.

Of course, I don't wish to alienate people who agree with me on most things. However, there are people on my side of the debate who, in an effort to fight fire with fire, will portray things in a risky way. The sad fact is that anything we say can and will be twisted and used against us. From the piece:

> I want to offer a gentle reminder to people on my side of the global warming issue who are promoting the idea that the drop in global

temperatures is the biggest on record in the satellite era.

1) The reason it's such a big drop *is because it was so warm.*

2) This was easily seen — it was forecasted in 2015.

The peak was so outrageous before that the drop after is huge, but it's not as great yet against the means as previous transitions from El Niño to La Niña. And guess what? This La Niña is shot. However, the bigger fish to fry is the Indian Ocean Dipole, the Pacific Decadol Oscillation and the Atlantic Multidecadol Oscillation. Like the '60s and '70s, these are due to turn cold (or at least less warm), which would mean cooling perhaps back to where temperatures were in the late 1970s.

Let's make sure everything is put in perspective. Biggest drop ever? Maybe. But it would not be if it wasn't the warmest we have measured in the past 35 years. It's not like it was normal and dropped to -1°C.

The fact is that the warmer it gets, the harder it is to make it warmer and the more likely it is to cool. The question then becomes, are we establishing a new base that was higher than the last 20 years? That does not mean its CO_2-driven. But it's like the old Ross Perot argument. Bill Clinton was telling everyone how much he improved the economy in Arkansas. Perot countered by saying if you double a penny, it's only worth a penny more but still twice as much. A major drop being touted was only because of how warm it was before. What is wild is how warm it still was *after the drop.* As I have tried to get across, I look at data and have to go where it takes me.

There are many blogs I have written on sea ice. This is another one of the hysteria points where you can waste time forever countering claims. I wrote "Debunking the Arctic Death Spiral Myth" on June 1, 2017. It's like an annual event as every year the shrinking ice cap cries arise.

Ah yes, another summer. But even though record low Arctic ice is highly unlikely, a forecast that *is* a sure bet is that climate alarmists will let us know about how much the Arctic is melting. **[By the way, there was no record low summer ice — it was in the middle of the pack in relation to recent years.]**

While winter levels were indeed at record low levels for that time of the year, the rate of melt has slowed so much that we are now closer to the running 30-year mean than we were last year at this time.

Records of ice and Arctic temperatures can be found on the Danish Meteorological Center website, which also has records of the current snow and ice situation in Greenland (it's way above average for the date). The Danish Meteorological Institute has Arctic temperature records going back to 1958. I have gone through all of them and have concluded that summers in the Arctic are no warmer (and may be a little cooler the past few years) and winters are warmer.

The winter was very warm, but it turned colder than normal in the spring and stayed that way in the summer. There have been other warm winters — 1974 and 1976 had very impressive warm spikes—but even with spikes it remains well below freezing, like it did this past winter. When one looks at actual records, it leads to improved perspective.

Looking through the records over the last 60 years, the warming is not all that great, and what has occurred is in the heart of the coldest time of the year. But even when CO_2 was much lower, there are examples of warmth in the Arctic. The cry then—even though it's being erased and denied now — was global cooling! In fact, the warming in the winter means it's water vapor that has a big impact on temperature increases where it's very cold and dry. That is the key. Slight increases in water vapor lead to higher temperatures. It's not CO_2, which has been increasing steadily in minute amounts. A naturally warming ocean due to cyclical events adds more water vapor and it affects cold dry areas more. It also disproportionately affects nighttime lows, which is why it's not the maximum temperatures that are warming but the nighttime lows.

John Holdren, President Obama's science adviser, wrote these words in his book from the early 1970s, *Global Ecology: Readings Toward a Rational Strategy for Man*:

"It seems ... that a competing effect has dominated the situation since 1940. This is the reduced transparency of the atmosphere to incoming light as a result of urban air pollution (smoke, aerosols), agricultural air pollution (dust), and volcanic ash. This screening phenomenon is said to be responsible for the present world cooling trend—a total of about .2°C in the world mean surface temperature over the past quarter century. **This number seems small until it is realized that a decrease of only 4°C would probably be sufficient to start another ice age.** Moreover, other effects besides simple screening by air pollution **threaten to move us in the same direction.** In particular, a mere **one percent increase in low cloud cover would decrease the surface temperature by .8°C.** We may be in the process of **providing just such a cloud increase, and more, by adding man-made condensation nuclei to the atmosphere in the form of jet exhausts** and other suitable pollutants. **A final push in the cooling direction comes from man-made changes in the direct reflectivity of the earth's surface** (albedo) through urbanization, deforestation, and the enlargement of deserts. **The effects of a new ice age on agriculture and the supportability of large human populations scarcely need elaboration here.** Even more dramatic results are possible, however; for instance, a sudden outward slumping in the Antarctic ice cap, induced by added weight, could generate **a tidal wave of proportions unprecedented in recorded history...**"

Notice the wording to encourage rapid action on this matter. Unprecedented in recorded history... Sound familiar?

Here's what's interesting. Volcanic ash is natural. But in regard to the other man-made causes, there is no mention of CO_2—unless one wishes to argue that something you exhale 100 times more than you inhale and that plants use to grow is a pollutant. (That is a key contention from modern day climate alarmists—to label a colorless, odorless gas needed for life a pollutant. It wasn't then.) In fact, we have really done great in this country getting rid of urban air pollution and agricultural air pollution.

Since Holdren's idea is that these are causes of *cooling*, would not the opposite be true—if we took care of them it would lead to warming? Look at his comments about urbanization, deforestation and enlargements of deserts. All of these are man-made, and none of them had anything to do with CO_2. But he was using them as a rationale for cooling when they are perhaps *actual arguments for warming!*

Another irony: The earth is getting greener, but if deforestation makes it cooler, as Holdren claims, is a greener planet then making it warmer? So by Holdren's logic, we have two positive effects—the cleaning of the air and more greenery that would also lead to warming!

The bottom line is people like this believe that anything you do to create a better world leads to adversity. Clean the air, it warms the planet. Pollute the air, it cools the globe. But wait— isn't CO_2 a pollutant, therefore it's now helping to warm our world? CO_2 helps plants! More plants, warmer earth. Up is down, down is up.

Urbanization is huge. Holdren thinks this leads to cooling? Does anyone really believe that? We had articles about the urban heat island from the '50s and '60s, but Holdren thinks it leads to cooling here? Think about what has happened around the big cities over the last 40 years and how thermometers that were relatively exposed are now surrounded by concrete. Las Vegas is a classic example—the airport there was in the middle of nowhere 50 years ago; now it's surrounded by sprawl, so it warmed. Even at Penn State, which sits in the middle of rural Pennsylvania, the site of the thermometer when I left school in 1978 was exposed to a golf course to the southwest, so around sunrise, chilly air would drain in on calm nights. Now it's surrounded by buildings, including one that bridges across US 322 and blocks any cool air drainage from areas where it's much chillier at night.

Urbanization and many other man-made effects have very real influences that lead to warming! The point is that nowhere does Holdren mention CO_2 and, ironically, he even had causes for cooling listed that if cleaned up —as they were—would have to result in the opposite by his own logic! And we have taken steps to counter them, so by his logic, would not the steps we took to improve those items have led to warming that the climate alarmist crowd is blaming on something else?

This was President Obama's science adviser. And I hear people complaining about Dr. Will Happer, arguably the world's foremost nuclear physicist, being unqualified?

What is the conclusion? I wouldn't panic about the Arctic melting until such a time when summer temperatures are higher by more than a degree or two. And there is no sign of that. If you want a simple example, just try to defrost a freezer if it's encased in ice. Now think about how long that takes (even with me using a hot blow drier to move things along) and imagine ice multi-feet thick in air that is slightly above freezing during summer. I wouldn't sweat the ice cap, and given U.S.-model-based CFSv2 anomaly forecasts (they are likely overdone, but they certainly are not crashing it to record lows), Arctic temperatures are turning lower than normal going into the warm season.

The fact that climate alarmists claim any answer as their own has been talked about, but here is one that had a lot of people talking: "First It Was No Snow and Cold. Now It's More Snow and Cold?" written on Nov. 1, 2016.

I will keep this short. The climate change (a.k.a. global warming) alarmists are now understanding that blocking over the North Pole is the inevitable result of long-term oceanic and solar fluctuations (or at least I hope they are). However, in an effort to promote their missive (fear of a cold, snowy winter), they are preemptively taking credit for supposedly anticipating a polar vortex shift. This is rich because during the winters of the 1970s, when it got very warm relative to average over the poles, we had people warning that an ice age was coming (of course they want you to forget that).

In the winters of 2013-2014 and 2014-2015 we had a lot of blocking similar to the winters in the ice age scare years of the 1970s. Blocking is in essence caused by it being warm over and around the polar regions. The fact is that this has been happening forever. It's cyclical in nature. Take, for example, the blowtorch winters of 2001-2002 and 2011-2012. The temperature anomalies were reversed. When cold congregates at the pole, the U.S. warms.

In the decadol sense, we are close to the late 1950s and 1960s with a similar sea surface temperature configuration. Guess what went on in those winters and is likely again this winter?

It's true the globe is warmer now, and that is where the real debate should focus. Nobody should take what is natural and make it something it is not. But it's also true there were articles claiming less snow and less cold were on the way. The fact is that the entire system is designed for conflict, and conflict is oftentimes resolved through extreme swings. We can observe it now and argue over the cause or what has the greatest influence (in the non-climate alarmist community, the arguments always rage between solar and oceans), but I guess here is the question one must ask alarmists:

Given the record of statements back and forth on this matter, *just what answer is not yours?* Is anything that happens a sign you are right? If so, why is anyone paying you since no matter what happens there is just one answer?

You can see clear links, in this case to the past. If it's happened before, it'll happen again. Sure, there are variations, sometimes warmer and sometimes colder, but it seems to me that a lot of the things that come out *rely on the general public* not being aware of history. In fact, that is applicable to many issues nowadays.

Now ponder this cyclical climate change theory: If it is colder and snowier farther south, would that not in turn over the long run start to cool the Arctic? After all, with more cold over the continent hanging around later in the spring and then starting earlier in the fall, as we have seen in the last several Eurasian winters, would that not begin to take its toll? And what happens when the oceans, which have been in the warm cycles in tandem, flip cold? Questions, questions, questions, and not a single one involves CO_2, though as an open-minded observer I cannot eliminate it having some effect. But how much? Apparently to many pushing this, it's now the climate control knob.

Let me see if you can make sense out of this. Is it going to snow in Detroit this winter? Your answer would be yes, right? Why? Because it snows every winter in Detroit. Now suppose I said to you, "No, that's not the reason. The reason is because of something different." Would you believe that? Of course not. Why? Because it snows every winter in Detroit.

The earth is warming now. It has warmed and cooled before. It always has. So why is it that CO_2 is now causing it when every other time before it was not?

There is nothing new under the sun. Perhaps you're just finally seeing some of what there is to see.

One of the biggest problems I have with the whole climate change agenda has to do with the fact we have much bigger problems to deal with now. It's arrogant and egotistical to think you *know* the answer about future events. You may have a pretty good idea, and when it occurs it may prove you right. But the problems that are occurring now are far more important. In fact, I think one of the reasons people love to obsess over the climate issue is because they don't have to deal with the hard problems — like people starving in our streets, or the decay of our cities, or our culture. So they go after a problem they cannot be judged for since it's so far down the road. What's wild is that many of the policies being advocated by way of the climate agenda *worsen* the problems we have now simply by deflecting money and attention from them. I took on the idea of taking all this climate change money and funding pre-existing conditions with it.

"Defund Climate Change Hysteria to Pay for Pre-Existing Conditions" was written on May 8, 2017, and addresses the typical utopian argument whereby we fight a straw man with no stated goal and intertwine it with the very real problem of health care.

Here's a novel idea.

Take the billions of dollars that are going toward what supposedly is a settled science issue — climate change — and use it to create a pool for pre-existing conditions. It is our duty to help those less fortunate and for the government to provide a safety net. So let's form that safety net by *dealing with a known problem today*, not a ghost that may or may not be there tomorrow — especially since human progress has skyrocketed in the age of fossil fuels. Do you think medicine would be where it is today absent the fossil fuel era?

The rest of the nation would be in the free market for insurance. And combined with tort reform and portability, we may be able to bring the price down.

We can't run from the problems of today, nor can we run from the record of the past. People are much more valuable than a few molecules of CO_2.

I doubt the American people approve of billions of dollars being spent on researching whether or not the earth is flat (no offense to Kyrie Irving) or other forms of "settled science." So for the sake of those suffering from pre-existing conditions, why don't we take the grant money allocated for climate change research and give it to those who really need it? If it's "settled science," then give up the money. *You can't have it both ways!* What about investing in our inner cities? Do climate change researchers need the money more than our sick, poor and needy? I think not. I know not.

This piece also went viral: "The March for … What?" published on April 21, 2017. This was in response to the ultimate march against a straw man by people aspiring to be heroic in a struggle that isn't really there. The usual suspects showed up (I was not one of them).

The March for Science is tomorrow and no one in their right mind would say they are against it because of its name. First of all, you are standing against the right of people to march for whatever cause they wish. Second, you would be portrayed as someone who is against science.

I am all for science. I think the climate changes. It always has and always will. Yet I have been portrayed as anti-science and a climate change "denier" by many who will be marching for things I certainly believe in.

Just who does not believe in science? It's a straw man that the marchers are marching against.

What's questionable is the way science is being portrayed and used. There is no apparent linkage between CO_2 and temperature in a time scale that goes back millions of years. So as someone who is acquainted with the scientific method, I am instantly skeptical of the idea that after all this time, there is now a linkage. *That does not mean there can't be,* and I am open to that argument and understand it. But as I asked in my last blog, how much linkage is there?

What I am trying to figure out is why there is a march when many of the people in that march have no tolerance for the questioning of their position. While I think it's noble to be inclusive and diverse, are any "skeptics" included as speakers? Is there diversity of thought? Of course not. Because in spite of what you see in countless charts that counter their ideas, they ignore the obvious. *The planet has always had temperature swings* — larger than today's and independent of CO_2—that should make any person searching for the truth skeptical as to how much CO_2 contributes.

Questioning of dogma need not apply. That sounds more like religion than science. Being for science means being for discussion. So who is anti-science here? This is a classic case of "blame your opposition for what you are actually doing." It is not the skeptic side shutting down debate.

One must be very careful when questioning the motives in academia. There seems to be two opposing forces today in society in general: people who seek to earn their keep, and people who believe they are owed their keep. There is no question that without research we would not be where we are today. But guess what fuels the economic engine that allows people the grant money, etc., for research?

I have to question motivation. For instance, if man-made global warming is such a done deal, why are we researching it anymore? Actual settled science (freezing and boiling points of water, gravity, the sun is darn hot) is not being researched. So apparently AGW is not settled science. And for good reason—if it is true this is all man-made, it's the first time, *established by science*, in recorded history. Another reason for being skeptical.

But the statement by the former EPA director that the actions have shut down a lot of business in this country and were brakes on the American economic engine really says a lot about what may be behind this. Preventing warming of only .01°C (you can't even measure that with certainty) over 30 years was not the main reason. Instead, it was to be a good example for the rest of the world. When I heard that it was so absurd to me I thought it was meant to sabotage the EPA mission. But no one said boo about it.

Finally, there seems to be a mass denial (there is that nasty word) that the progress of humans, and of course researchers, has been huge in the fossil fuel era.

The assumption that this would not continue makes no sense. In addition, a vibrant economy seems to be a moral and ethical positive. As far as researchers worried about grants being cut, would you rather get 10 percent of 50 or 15 percent of 10? Yes, it's a bit of an exaggeration, though it makes my point. The population curve and the increase in GDP and life expectancy says to me the pie is expanding and many new challenges that need researching are going to continue to challenge us. And science will have to meet that challenge.

I will not be going to the March for Science. I rather doubt I would be welcome. I would have to go disguised for fear of being torn limb from limb by the open, tolerant marchers. But like all questions in science, when I look at the march, I am asking why they're marching.

No one is anti-science, even if a group of people try to convince you otherwise.

These varied blogs apparently attracted the most attention. I never know beforehand if something is going to get a lot of attention. My approach to this issue is not to look for something to write about but to write on a matter when it comes to me. It's like making a record — while some can just write a three-minute hit at will with a formula, there are others who need to be inspired. I guess I am like that. And like a lot of music, there are varied tastes. I always take a look at the amount of people who "recommend" a piece after it goes public, partly because I am competitive and partly because I am curious as to what people thought. In any case, I thought I would share the top five here and comment on them.

Chapter 9

In God We Trust — or Do We?

It is amazing that in a nation based on Judeo-Christian principles and a constitution inspired by them, no one can talk about God or politics anymore without the wrath of someone coming down on them. In the climate debate, I have a simple way of judging things in this matter.

In relation to God, are you using God to further your aim? For instance, the idea that Jesus would have loved Communism. For someone like me, that is a non-starter.

In relation to government, are you using the climate issue to further your political aims? That is a non-starter.

We have discussed the latter, so briefly I will show what I have said on the former.

My attitude is not dissimilar to Pastor Eric Liddle in "Chariots of Fire." There is a scene where he says, "God made me fast and I feel His pleasure when I run." For me, God gave me weather and climate and I see his majesty in them every day. I use weather and climate as instruments to get closer to God. I don't use God to further my goals. What this means is that I am in awe of Creation, which, of course, may blind me to understanding why the increase of one molecule of CO_2 out of every 10,000 molecules of air over 100 years is going to thwart the sun, the oceans, stochastic events and a system that is designed for conflict and resolution. Reduce CO_2 by 25 percent and it's no big deal. Reduce incoming radiation by 25 percent (or even .25 percent) and it's bye-bye. Nature thrives on correcting imbalances that are its natural challenges, just as the challenges placed in front of man has led to his advancement. But the challenges that spurred prosperity were attainable. Trying to control nature and pretending you are the master of Creation set man up as god. Because if man can control what God created, is not man greater than God? If that is the case, then men who see themselves as god seek to control those under them.

Look at this passage from the book *The Ragamuffin Gospel* by Brendon Manning, from the site afterall.net.

To Quote Brennan Manning in his consideration of "The Fine-Tuned Universe":

"The slant of the earth, for example, tilted at an angle at 23 degrees, produces our season,. Scientists tell us that if the earth had not been tilted exactly as it is, vapors from the oceans would move both north and south, piling up continents of ice. If the moon were only 50,000 miles away from earth instead of 200,000 the tides might be so enormous that all continents would be submerged in water, even the mountains would be eroded. If the crust of the earth had been only ten feet thicker, there would be no oxygen, and without it all animal life would die. Had the oceans been a few feet deeper, carbon dioxide and oxygen would have absorbed and no vegetable life would exist. The earth's weight has been estimated at six sextillion tons (that's a six with 21 zeros). Yet it is perfectly balanced and turns easily on its axis. It revolves daily at the rate of more than 1,000 miles per hour or 25,000 miles each day. This adds up to nine million miles a year. Considering the tremendous weight of six sextillion tons rolling at this fantastic speed around an invisible axis, held in place by unseen bands of gravitation, the words of Job 26:7 take on unparalleled significance: 'He poised the earth on nothingness.' The earth revolves in its own orbit around the sun, making the long elliptical circuit of six hundred million miles each year — which means we are traveling in orbit at 19 miles per second or 1,140 miles per hour. Job further invites us to meditate on 'the wonders of God' (37:14). Consider the sun. Every square yard of the sun's surface is emitting constantly an energy level of 130,000 horse power (that is, approximately 450 eight-cylinder automobile engines), in flames that are being produced by an energy source much more powerful than coal. The nine major planets in our solar system range in distance from the sun from 36 million to about 3 trillion, 6,664 billion miles; yet each moves around the sun in exact precision, with orbits ranging from 88 days for Mercury to 248 years for Pluto. Still, the sun is only one minor star in the 100 billion orbs which comprise our Milky Way galaxy. if you were to hold out a dime, a ten-cent piece, at arm's length, the coin would block out 15 million stars from your view, if your eyes could see with that power."

Startling.

Of course, one way to avoid that argument is to take the stand that there is no God. But even then, if you can control nature, are you not a god?

Seems arrogant to me.

Furthermore, if you believe in God, why would He create an organism that exhales 100 times more CO_2 than it inhales and command it to be fruitful and multiply? Would it not self-destruct? The same God created plant life that thrives on CO_2 and is the source of fossil fuels anyway. A majestic synergism has been set up.

If you don't believe in God, that's fine. But you have to admit it's one heck of a grand coincidence that animals and plants are tied together in a way in which they mutually benefit each other.

I don't delve into this relationship too much. God and nature are infinite, and man is finite. In the comprehension of God by man, the only bridge can be faith, since the finite is infinitely smaller than the infinite. I am not fit to judge anyone, so if someone does not believe in God, that is his or her choice. It's when that person sets him or herself up as a god, claims to know the future and then seeks to enforce his or her will on another person that raises red flags. I don't think that is an unreasonable position.

I did delve into challenging a spiritual authority on one blog, "A Quick Note to the Pope," written on Sept. 18, 2015. In the piece I express alarm over the pope's stance on climate change when the human condition has flourished since the advent of fossil-fueled technology.

It was short and to the point. Many in the anti-climate change community love this pope. It's funny because I grew up Catholic. I did not see them endorsing his position, for instance, on the veneration of the Virgin Mary and basically every single other thing he stands for. But what is fascinating is how climate alarmists use the pope on one issue if it suits them, even though they have little common ground on every other issue. And by wading in here, the pope has one foot in the city of man and one in the city of God. A sure way to get torn up as history has shown.

In any case, I don't believe we should be trashing nature. But invoking God (if you believe that) when He created a symbiotic relationship between animals and plants (carbon dioxide being the common link) as a reason CO_2 is a life-threatening pollutant makes no sense. I see God's hand in nature and I am thankful for it. I don't believe in using God as a weapon to implement policies that seem to go against the very nature of His creation.

It comes down to the absolute truths of nature and nature's God versus the relative truth of men who hold the type of motives you see. It's damming evidence of motives that have nothing to do with my reason for writing this book, which is strictly the climate and weather.

It is a strange world when the actual words of people are hidden or discounted. Deception in any form is certainly a tool of evil. Are you going to trust humans with despicable motives? If you don't believe in God, again, I am not out to change you. I am merely pointing out that when you love something, you seek to find the right answer. If you don't believe in God, well, who do you trust? The more you know, the more the evidence should make you skeptical of the climate change agenda.

The pope weighed in again in 2017, labeling people who do not subscribe to climate change worship as "perverse."

I wrote the following, but it was not published.

So Now I Am Perverse?

I've been thinking about this San Francisco Chronicle headline: "Pope rebukes climate deniers as 'perverse' in Bonn message."

When Marxism creeps into spiritual matters, this is the result. I guess the millions of years of records that show no CO_2-temperature relationship is now considered perverse in the eyes of the pope.

So how does the pope explain the fact that life has thrived when it's warm and is thriving and progressing more than ever because of the fossil fuel era? Per capita income, life expectancy and the amount of people have all skyrocketed. It seems that the fossil fuel era has helped the church to be fruitful and multiply. Man has certainly multiplied, and there are more people today enjoying the fruits of their labor than ever before.

The answer, of course, is he can't explain any of this. He is intellectually and spiritually bankrupt on this matter, blinded by a Marxist takeover of Judeo-Christian beliefs that formed the church. Perhaps he should quit trying to "reform" the church and attempt instead to restore it to its original foundation. It would give him a clearer vision. But that's what arrogance does. It rewrites and eliminates knowns to pursue some unknown of its own creation. One problem — if man is creating it, it's not the same as God creating it.

And if you don't believe in God, fine, but you can't deny the progress of the human condition in the fossil fuel era or the cyclical nature of past climates or climate optimums. Or is optimum a perverse thing now?

The bottom line of all this: The people who are promoting this agenda have no qualms about using God for their purpose. As someone who believes fervently in God, the use of God as a tool is demeaning and insulting, something I am sure does not bother climate alarmists in the least. If you do not believe in God, then why would you put faith in something you do not believe in anyway?

If you are a man of God and a leader of people reaching for God, having one foot in the city of man, which is clearly what the climate issue is, and one foot in the city of God will get you torn apart, along with your church for that matter.

This from a man who comes from a Catholic background.

Chapter 10

Conclusion

I said in the introduction that this is a love story. It Is. *The Climate Chronicles: Inconvenient Revelations You Won't Hear From Al Gore—And Others* is a wandering compilation of blogs I have written because something I love is being strewn to the four winds and ruined. When you understand my motivation, you see that I am not out to save the world or be someone I am not. The fact is you must decide for yourself what you believe about climate change. I have explained where I stand and why, but you must evaluate your own intuition. They call me a denier. Of what? That the climate changes? Of course it changes, and up until now CO_2 was never the climate control knob. But they do it to isolate, demonize and destroy. As for my reference to Climate Ambulance Chasers, well, it's tongue in cheek, but I do think there is some truth to the comparisons. More than saying I am a climate denier (just what is that?) because I question whether CO_2 is now the control knob of the climate when it never has been.

I love my country. I feel that this issue is being used to demonize and destroy our way of life. That is the real purpose of it all. So, in a way, this book is a defense of our nation.

I believe that the foundation you stand on today was built on the structure of the past and will help you reach for the future. I believe this is being attacked also; our foundational truths are being destroyed. So, in a way, this book is a defense of our foundation.

I believe there is a great struggle today between the absolute truth of nature and nature's God and that men today, in their increased wisdom and power, seek to play god by claiming they know the future and that they possess the wisdom to force others to obey. So this book is a defense of nature and nature's God.

You don't have to believe in God. That is your choice. But certainly that excludes other men no greater than you being God like and forcing you to adhere to their ideas.
The fight over weather and climate is really not what this debate is about. It is simply a means to an end. An age-old tool of deception.

The fact is, I am seeing things today that I have not seen before, and not in a way that screams progress. Arguments that are put forth ignore common sense (i.e., if it happened before, it can happen again). And the way it is being put forth is a different world from what I grew up in. You would have to become part of that world, risk everything that you believe and put your very being at risk to understand it. I wrote about it in "No Country for Old Weathermen" on July 24, 2013.

The movie "No Country for Old Men" is one of my favorites, as the Tommy Lee Jones character tries to cope with a world gone mad. Here's one of my favorite passages:

"You can't help but compare yourself against the old-timers. Can't help but wonder how they would've operated in these times.... I don't know what to make of that. I surely don't. ... [I]t's hard to even take its measure. It's not that I'm afraid of it. I always knew you had to be willing to die to even do this job—not to be glorious. But I don't want to push my chips forward and go out and meet something I don't understand. You can say it's my job to fight it but I don't know what it is anymore. More than that, I don't want to know. A man would have to put his soul at hazard. He would have to say, 'Okay, I'll be part of this world...'"

I know how he feels.

What I'm seeing today, and it goes beyond the climate fight, makes little sense to me. An agenda that once preached *global warming* has been turned into something that's now simply *climate* — the back-and-forth weather that has occurred naturally throughout the ages. Propagandists say that skeptics deny the existence of climate change — a redundant term — and then seek to isolate and demonize those of us who point out their fallacies. They are relying on the fact that most people never bothered to look at history. Guys like me make a living off researching the past to give us an advantage looking forward. Yet we have a whole group of people who have changed the rules of the game because they are getting beaten. (Remember, we were supposed to be beyond the tipping point by now because of man-made global warming.) They take events from the past that many may not know about, pull out examples of it happening today that everyone can see in an instant, and claim it's because of anthropogenic "climate change."

I can give example after example, but here's one using global sea ice. You know all this screaming about the Arctic ice cap? Consider these points

- It's done at the expense of a record *high* Antarctic ice cap just a few years ago.

- How about this inconvenient fact: The Arctic had more ice melt at the end of the last warm cycle of the Atlantic (our presently warm AMO should end in several years).

We have pictures of submarines surfaced at the North Pole, one of them from the '60s.

If the Pole had little ice around it like it did in 1962, you can be sure it would be plastered all over every mainstream paper and environmental blog. But would they tell you that it has happened before and should actually happen again in this warm cycle?

Or how about this: Sea ice records conveniently started on most charts when it was at a peak in the late 1970s. If we had started in the early '70s, there was much less.

I love the movie "No Country for Old Men" because there is a real message that pertains here. In all the craziness you have to stay true to what you know and love. I have loved the weather since the first day of my memory. I know the climate because of my love for the weather and my effort to find out what drives it. It's like the study of history: Find out why things happen and apply them to tomorrow. I will always have the weather, even when this fight is over, and that drives me. But if your goal is to save the planet, or to force your way of doing things on someone else, or if everything you have done is based on the idea that man is causing the climate to change, what happens if you are found to be wrong? What do you have? What happens when your belief—that only through the collective can there be progress—is challenged by people who cherish individuality and believe that's the best way? The individual does not wish to limit other people, and it doesn't threaten his or her beliefs that they can do it on their own. But someone who believes everyone must be on board is threatened by the idea that someone can be right on their own. And that may be a reason for all the craziness we see. It leads to a huge difference in motive.

The end of the movie, a second dream recollection, hit me like a ton of bricks:

"But the second one it was like we was both back in older times and I was on horseback going through the mountains of a night. Going through this pass in the mountains. It was cold and there was snow on the ground and he rode past me and kept on going. Never said nothing. He just rode on past and he had this blanket wrapped around him and he had his head down and when he rode past I seen he was carrying fire in a horn the way people used to do and I could see the horn from the light inside of it. About the color of the moon. And in the dream I knew that he was going on ahead and that he was fixing to make a fire somewhere out there in all that dark and all that cold and I knew that whenever I got there he would be there. And then I woke up…"

That somehow the light from what came before will guide us in the future provides some comfort to me. But what happens if that light is hidden or there are people seeking to put it out? I guess the only thing left to do is to fight without losing your soul.

The ultimate irony: Tommy Lee Jones was Al Gore's roommate for a time in college.

Above all, this is a love story. But weather and climate and my love of them are only a part of it. And in a world gone mad, perhaps it's only our grasp and defense of what we love that makes sense.

About the Author

Joe Bastardi has been referred to as an institution in the science of weather prediction. Many companies across a multitude of industries, from energy to retail, have profited from his forecasts. His exceptional skills are rooted in a comprehensive understanding of global oscillations and in-depth analysis of historical weather patterns.

This contributes to his skepticism of claims that are being made as to how bad things are now, since there are so many examples as bad or worse! The book brings out many of these and much more.

Appendix

A partial list of the articles I have read to formulate my ideas.

- https://wattsupwiththat.com/.../confessions-of-a-greenpeace.../...
- https://wattsupwiththat.com/.../nature-abhors-a-positive-fe.../...
- https://tallbloke.wordpress.com/.../decadal-lag-of-temperatu.../
- http://www.sepp.org/key.../Climate%20Fears%20and%20Finance.pdf
- https://wattsupwiththat.com/.../the-diminishing-influence-o.../...
- http://tropical.atmos.colostate.edu/.../Publicat.../gray2012.pdf
- http://thehill.com/.../214877-the-climate-change-money-machine
- http://hockeyschtick.blogspot.com/.../updated-list-of-29-excu...
- https://wattsupwiththat.com/.../amopdo-temperature-variation.../
- http://motls.blogspot.ca/.../le-chateliers-principle-and-natu...
- http://www.c3headlines.com/global-warming-quotes-climate-ch... (This one is downright ugly. It shows the kind of people we are dealing with.)
- http://www.crisismagazine.com/.../mixing-up-the-sciences-of-h...
- https://wattsupwiththat.com/.../22-very-inconvenient-climate.../
- https://wattsupwiththat.com/.../the-climate-wars-damage-to-s.../
- https://wattsupwiththat.com/.../the-mathematics-of-

carbon-di…/

- http://sciencespeak.com/climate-basic.html
- http://thefederalistpapers.org/…/top-10-climate-change-pred…
- https://wattsupwiththat.com/…/five-points-about-climate-ch…/
- https://www.youtube.com/watch?v=qxTmM9dDjrY&feature=youtu.be
- http://joannenova.com.au/…/forgotten-extreme-heat-el-nino-…/
- http://climatechangedispatch.com/how-to-rebut-climate-hyst…/
- http://notrickszone.com/…/theres-no-observational-physica…/…
- http://news.nationalpost.com/…/alex-epstein-wrapping-our-mi…
- http://www.breitbart.com/…/political-science-reply-375-con…/
- http://notrickszone.com/…/17-new-2017-scientific-papers-a…/…
- http://www.dailywire.com/…/scientists-we-know-what-really-c…
- https://wattsupwiththat.com/2014/08/10/the-diminishing-influence-of-increasing-carbon-dioxide-on-temperature/

Made in the USA
Las Vegas, NV
29 September 2022

56165951R00125